HISTORY
of the
WATER POLLUTION CONTROL FEDERATION
1928-1977

Dedication

This history of the Water Pollution Control Federation is dedicated to the thousands of individuals who, over the past 50 years, played a part in the art of water pollution control, even though they may not have been members of the Federation. Without those efforts, this History would be little more than a digest of meetings of a group of people seemingly with the same interest.

Committee on Federation History

    Sidney A. Berkowitz
    Robert A. Canham
    Ralph E. Fuhrman
    Paul D. Haney
    Harris F. Seidel
    William H. Wisely
    George E. Symons, Chairman

Water Pollution Control Federation
2626 Pennsylvania Avenue, N.W.
Washington, D.C. 20037

Copyright © 1977 by the Water Pollution Control Federation,
Washington, D.C. 20037 U.S.A.
Library of Congress Catalog No. 65-5192

Printed in U.S.A., 1977
by
Moore & Moore, Inc., Washington, D.C.

Preface

The Water Pollution Control Federation, as it exists today, was a child of adversity—born on the eve of the worst national economic depression in history, and reorganized just before the U.S. entered World War II. The fact that the Federation survived the depression—even so narrowly—is proof that there was at the time a need for communication and interaction among the few technologists who chose to mold their careers on the obvious necessity for attention to water pollution problems. Most remarkable, however, was the development and solid growth that occurred during World War II and the decades thereafter.

At the time the Federation was organized, less than thirty percent of the urban sewage of the U.S. was treated at all, and few of these treatment facilities were highly effective. Raw sewage flowed freely into rivers, lakes, and marine waters. At the time of the 1941 expansion and reorganization of the Federation, there were 5,600 municipal treatment plants serving half of the sewered population; by 1955, more than 7,500 plants were treating over 75 percent of the wastewater produced by the cities of America.

This progress was made in the face of public apathy—even resistance—to sewage works projects. Public financing was difficult, and shortages of critical materials were common. It was state legislation and action that provided the original impetus to the abatement of water pollution, and a few hundred dedicated specialists provided the know-how. Through the medium of their fledgeling organization they advanced the management of water pollution from an empirical art to a sophisticated spectrum of science. Indeed, the Federation was very much a part of this pioneering action.

As the 50th anniversary of the Federation approaches, it is fitting that this history be assembled. The idea for a history of the Federation was first broached by Paul Haney, then Federation President-Elect, in June, 1968. Mr. Haney proposed the idea officially in July of that year and the Board of Control approved the appointment of the *ad hoc* Committee Chairman, George E. Symons, at the Chicago WPCF Conference in October, 1968.

Selection of the Committee was left to the Chairman, who chose Sidney Berkowitz, Robert Canham, Ralph Fuhrman, Paul Haney, Harris Seidel, and William H. Wisely. The Committee is indebted to George Burke, Jr. of the Federation Headquarters staff for supplying much of the detailed statistical data that was buried in old minutes, reports, and journals.

At the Chicago Conference of the Federation in 1968, Messrs. Berkowitz, Fuhrman, Haney, and Symons decided that a letter should be dispatched to a number of persons who could contribute historical information. At that time it was determined to collect information and organize it into six general categories: (1) organizations that existed before the Federation was organized in 1928; (2) activities of the Committee of One Hundred that organized the Federation, then known as the Federation of Sewage Works Associations; (3) organization of those associations that became charter members of the Federation; (4) operations during the period 1928 to 1940 (Secretary H.E. Moses); (5) operations between 1941-1954 (Secretary William H. Wisely); and (6) operations during the period 1955 to date (Secretary R.E. Fuhrman). However, Dr. Fuhrman left the Federation at the end of June in 1969 and a fourth period was added to cover the tenure by R.A. Canham as Executive Secretary.

The Committee asked for information, data, publications, reports, and correspondence that applies to the above categories; any personal recollections that would be helpful in compiling this history of the Federation; and any suggestions as to other persons who should be queried.

The request for information was sent to Prof. Harold E. Babbitt, Earnest Boyce, E. Sherman Chase, Morris M. Cohn, Prof. Gordon M. Fair, F. Wellington Gilcreas, Dr. W.D. Hatfield, Arthur M. Niles, Walter A. Lyon, Prof. George M. Schroepfer, Robert A. Shaw, Walter A. Sperry, Henry M. Van der Vliet, and of course, to the Committee Members. Several of these persons were Past Presidents, or had worked with persons active in the early years of the Federation. Messrs. Babbitt, Chase, Cohn, Fair, and Niles have died since work began on the history. Harris Seidel, Robert Shaw and Walter Sperry supplied a wealth of information.

T.M. Niles was queried and reported that the files relating to early activities of Samuel Greeley had disappeared. Walter Lyon of the Pennsylvania State Health Department reported that Ted Moses' files had also disappeared. Some time later, at a meeting of the Pennsylvania Water Pollution Control Association, Deacon Matter spoke on the early years of the Pennsylvania Health Department. His reminiscences were taped, later transcribed, and appear in the Appendix to this History. Other personal recollections, particularly those of H.E. Babbitt, Earnest Boyce, Sherman Chase, Gordon Fair, and F. Wellington Gilcreas, also appear in the Appendix, as do the reminiscences of the committee members.

The assembly of the History began too late to include the recollections of many persons who had been active in the early efforts to establish the Federation; many of them had passed away.

The Committee envisioned a four-year program to produce a text ready for publication. Work on collection of material began in late 1968 and proceeded with some diligence for the next two years. Unfortunately, progress on the work suffered during the next two years because of other commitments of the Committee Chairman. Then came the change in Executive Secretaries and a decision was made by the committee to delay final publication until mid-1976, in order to make the history cover the Federation's first 50 years. It then was logical to plan the issuance at the 50th anniversary meeting scheduled for Philadelphia in October, 1977.

As with any history, no small group of persons can assemble all of the information that must exist in various files all over this continent and in foreign lands. It is hoped, however, that this volume will provide the framework for an expanded history for the 75th Anniversary in 2002.

Much of this history of the Water Pollution Control Federation is about people, because from its beginning the Federation's very existence revolved around individuals and their ideas, aspirations, efforts, and interest in the field of water pollution abatement. The men who are named in this book are a large part of the Federation's history—without them, there would have been no Water Pollution Control Federation.

October 1976

CONTENTS

Preface v

Chap. I   Genesis 1

    II   Emerson/Moses Era—1928-40 19

    III   The Coming of Age—1941-54 31

    IV   Headquarters Moved to Washington—
               1955-69 49

    V   Rounding Out a Half Century—
               1969-77 65

    VI   Federation Programs 99

    VII   Charter Association Histories 165

    VIII   Personal Reminiscences 197

         Appendix 261

Chapter One

Genesis

The ideas and efforts of several individuals originally led to the formation of an organization devoted to the needs of the sewage works operator. One of these early efforts was the formation of the Committee of One Hundred, which laid the groundwork for an organization informally designated as the "Sewage Works Association." As Harris Seidel* has written, "There was really no single time or place of the Federation's birth; nor was there any individual, or even several, of whom it might be said: 'They organized the Federation.'" There are, however, a number of people who contributed to its beginning, and George Warren Fuller appears to have made the first concrete formulations of the organization's concept.

In the mid-1920's, there were several sewage works operators' associations and short schools in existence, scattered through the South, the East, and the Midwest. A number of state health departments had sanitary engineering divisions and there were many well-designed sewage treatment plants in operation. Many of these, especially the larger plants, had technically trained superintendents. These groups and agencies were struggling along independently, with little exchange of ideas and very little in print to help operators with their practical problems.

Publication in the sewage works field also presented a problem. Research results and major plant descriptions found a limited outlet through the journals of the American Society of Civil Engineers, the American Chemical Society, the American Public Health Association, and sometimes the American Water Works Association. But very few persons who would benefit from a new organization could qualify or were interested in membership in those organizations. The commercial, municipal, and public works magazines of the day carried plant descriptions and stories on operation, but researchers found that these editors were not interested in the technical details of their work. Much of the early motivation for founding the Federation came from these researchers and others with something to say who needed a better place to say it.

---

* Portions of this chapter are based on material that Federation Past President Seidel prepared for a history of the Iowa Water Pollution Control Association. Inasmuch as Dr. Seidel is a member of the Federation History Committee, his material is included in the text without quotation or other acknowledgement, but with thanks.

How the Federation Began

The initial drive toward founding an organization began at a meeting of the Conference of State Sanitary Engineers in Buffalo, N.Y., on October 14, 1926. Charles A. Emerson, first President of the Federation, later commented* that at that meeting "it was decided to invite some interested outsiders to sit in on a general discussion. George W. Fuller, Dr. Willem Rudolfs, Arthur Bedell, and I were among the guests. Dr. Rudolfs made a plea for a better medium for publication of research papers. Others also spoke, but nothing tangible was accomplished. Throughout the meeting, George W. Fuller sat quietly smoking his pipe, but made no comments. However, at dinner at the hotel that evening, he said: 'I think I've got the answer and will tell you about it in a couple of days.'

"Shortly thereafter, Kenneth Allen, William J. Orchard, and I were called to his office. It was then and there that the Federation was conceived by keen-thinking, practical-minded George W. Fuller.

"His thoughts were:

"1. A national sewage association with an annual meeting was out of the question at the start, because few operators could obtain authorization to attend a convention which might be several hundred miles distant.

"2. Desired results could be obtained by strengthening existing state sewage works associations and formation of others. He felt that municipal officials would be inclined to authorize expenses for a trip by the operator to a meeting within his own state, particularly if requested to do so by their own State Department of Health.

"3. The local associations should be joined into a country-wide Federation, to be guided by representatives selected by each of the member groups to act as a Board of Control.

"4. A journal should be published and distributed to every member of the local associations at a moderate price.

---

* In "A Quarter Century of Progress," an address presented at the Silver Anniversary Luncheon, New York, N.Y., Oct. 8, 1952.

"Although the idea at once appealed to all of us, we were told to think it over and talk it over with friends and come back for another meeting in a week."

Emerson went on to say that "It developed at the following meeting, one week later, that Bill Orchard, with characteristic enthusiasm, had secured definite promises from 20 firms to take advertising space for one year in a sewerage journal. The contemplated advertising revenue, together with an estimated initial membership of 500 at $1.00 per year each, would make a total income of $7,200, or about enough for printing and distributing four issues of a journal of 100 pages each.

"Having this assurance of ability to finance a journal, the next step was to ascertain the probable support to be expected from health officials, consulting engineers, existing sewage works associations, and that most important group, the operators. Letters to key men brought enthusiastic replies; informal group meetings in Detroit, Cincinnati, and St. Louis also elicited many promises of support."

Early in 1927, at its meeting in New York City, the Sanitary Engineering Division of ASCE appointed a "Committee on Stream Pollution Matters" chaired by Harrison P. Eddy, with Langdon Pearse, Anson Marston, W.L. Stevenson, Richard Gould, Harry F. Ferguson, and W.H. Dittoe as committee members. According to Emerson, "Harrison P. Eddy sent letters to a selected group of sanitarians suggesting a dinner meeting in Chicago, on June 10, 1927, for discussion and formulation of plans to be presented at a general meeting to be held the following morning on adjournment of the AWWA Convention."

Apparently, Mr. Eddy was not the only one writing letters at the time, for Prof. Jack Hinman of the University of Iowa recalled some years later that, "Following a good deal of discussion at an Iowa sewage works meeting, it was decided that something ought to be done. After considering this further, I went ahead on my own and wrote a series of letters calling for a meeting to discuss this matter at the next AWWA Convention in Chicago: and meetings were held in June 1927 in the Sherman Hotel, I believe." In 1932, Dr. Max Levine credited Hinman for this letter before the 1927 AWWA meeting, although his name is not prominent in any activity prior to the formation of the Committee of One Hundred.

The ASCE Committee under Eddy summed up its efforts in a report at the ASCE national meeting in New York, N.Y. on January 28, 1928:

"Pursuant to the instructions of February, 1927, the Committee has endeavored to determine the policy and procedure which can best be adopted by the Executive Committee regarding increased facilities for presentation, discussion, and publication of data, results of research, and papers in the field of stream pollution, with particular reference to sewage and industrial wastes treatment and disposal.

"It was found that the American Water Works Association, the American Public Health Association, the Conference of State Sanitary Engineers, and the American Society of Municipal Improvements already had appointed committees which were giving more or less attention to this and allied subjects. The Committee arranged a conference for June 10, 1927, in Chicago, Ill., which was attended by fifty-seven persons interested in stream pollution. On the following day, a more formal meeting was held under the auspices of the American Water Works Association. Those conferences showed that there is a rather widespread interest in this subject and developed opinions indicating that:

"1. Many workers in this field are unable to travel long distances to attend conventions and, therefore, are deprived of opportunity for hearing and discussing papers.

"2. While there are several national organizations which deal with this general subject at conventions, there is little or no coordination of their activities, with the result that there is duplication of effort and material and the degree of efficiency is low.

"3. It is likely that sufficient valuable material will be available to provide manuscript for a journal of satisfactory size.

"4. The space available for such material in existing journals is so limited that it is frequently impossible to secure prompt publication.

"5. Facilities for the publication of papers reporting the results of research are not considered to be adequate by men working in this field."

The 1927 Meeting in Chicago

Engineering News Record in its June 16, 1927 issue, reported on the meeting following the AWWA Conference: "At a meeting held Saturday morning, under the auspices of a committee appointed by the executive committee to consider an outlet for papers on sewage research and the operation of sewage works, the *fifty or so present voted against* the formaton of a sewage section of the American Water Works Association, as reported in more detail, together with other discussions of this and related matters, elsewhere in these columns."

The other relevant "related matters" in that *ENR* issue read as follows:

Water Works Association Sticks to Water
Two Other Conferences Leave "Outlet"
for Sewage Works Data Unsettled

That the American Water Works Association has all it can handle in its own field and therefore will not form a sewage section was decided at its Chicago convention last week. Those wishing to consider some other means of co-ordinating and extending organized activities in the field of sewage-works operation and research met immediately after the sewage conference of the water-works association and voted to have the chairman of both meetings (C.A. Emerson, Jr., of Fuller & McClintock, Philadelphia, Pa.) appoint a committee on the subject, without instructions save to arrange for an open conference at the convention of the American Public Health Association next fall.

On the evening before the action just outlined, there was a dinner conference on the same general subject, called by H.P. Eddy, consulting engineer, Boston, Mass., chairman of a committee of the Sanitary Engineering Division of the American Society of Civil Engineers, created to consider the sewage activities of the division and their correlation with those of other organizations. Mr. Eddy called the dinner conference solely for the purposes of discussion —which was extensive but showed little unanimity of opinion. There were thus three sewage conferences at the AWWA convention, attended by substantially the same group of perhaps 60 sanitary engineers, chemists and biologists, nearly if not all members of the AWWA.

Basis of Decision

The decision not to form a sewage section in the AWWA was based on an analysis of the problems and finances of the association, outlined by Abel Wolman in behalf of a special committee created by the executive committee to consider a possible "outlet" for papers on sewage research and on the operation of sewage works. Mr. Wolman stated that the $18,000 budget allowance for the Journal is insufficient to print the papers on water works available; that large sums are in demand for the work of committees on steel pipe lines and boiler feed waters, the latter alone needing $50,000 in the next five or ten years. Moreover, although there are 10,000 water works in the United States and Canada, the association has only 2,500 members; and besides extending its membership in the water-works field, there is still much to do in the way of consolidation of water-works associations.

The inadvisability of AWWA expansion outside its own field having been set forth, Mr. Wolman reviewed other organizations which might go more extensively than now into sewage matters, but found that none of them could do so without embarrassment. He suggested a federation of the sewage activities of existing organizations, with occasional joint meetings rather than the creation of a new society; and the central collection and publication of sewage papers, throughout some existing technical society or some regular publisher.

Other speakers stated that the New Jersey Sewage Works Association had already voted to publish a quarterly open to other societies and urged the formation of still other state associations of sewage-works operators, and the affiliation of these and other societies wholly or partly concerned with sewage disposal.

In advocacy of a sewage-works section of the AWWA, I.W. Mendelsohn, sanitary engineer, U.S. Public Health Service, assumed that the many members of the Water Purification Division would all be interested in a sewage section, as also some 500 others he had checked off in the list of members, making 43 percent of the membership. He also presented the results of a letter canvass of the geographical sections, showing that 15 sections had favored a sewage-works section, 5 reported that the matter would be taken up later, and 7 had not yet replied.

C.M. Baker, engineer, Michigan Department of Health, in pleading for a broad plan of action, named 11 organizations which he thought should co-operate; American Association of Manufacturers, Izaak Walton League, ASCE, AWWA, ACS, American Biology Society, APHA, ASME, Conference of State Sanitation Engineers, U.S. Public Health Service and the U.S. Fisheries Bureau.

On being requested for an expression of opinion, Homer Calvert, secretary of APHA, and editor of its Journal suggested a careful survey to find how many pages on sewage-works lack a place for publication. Robert S. Weston, consulting engineer, Boston, Mass., said that a sewage section of AWWA would be an example of working in parallel, whereas he believed the better plan would be to work on the series system; that is keeping water and sewage matters separate. The final outcome of the three conferences was to leave no organization commited to any plan but, as already stated, to create an uninstructed committee of investigation, apart from either the AWWA or the committee of the Am.Soc.C.E. already in existence."

In that same issue, Engineering News-Record editorialized:

Federate and Concentrate

Prudent hesitancy in launching a new technical society was displayed by the group of men which last week considered the need for a new "outlet" for papers on sewage research, sewage-works operation and related matters. It was fortunate that this spirit guided the conclusions reached at the Chicago conference, for there is commonly too great a readiness to create new organizations in a field already well filled. There was little unanimity beyond this point, however. No agreement could be reached as to what should be done to deal with the situation that brought about the conference, although the feeling was clear that some instrumentality is needed which does not now exist. One of the best plans offered was that a federation of existing agencies be brought about, especially state organizations like the New Jersey Sewage Works Association. Such a plan would build from below instead of from above. It would supply co-operation and correlation, and prevent overlapping of national organizations. We venture to suggest that one or another of the national bodies might arrange a conference of all existing organizations dealing with sewage-works operation, in order to bring such federation about. There were those among the Chicago conferees, too, who would be satisfied with assembly in one publication of all the papers on sewage matters. This is another view worth considering. But obviously it should be preceded by a survey and a weighing of the need for a new "outlet"—as one speaker suggested. Even if there were to be a central publication, restriction rather than increase in the outflow would be welcomed by many, to make its bulk more manageable and raise its quality. From every point of view, thus, the best present policy is to federate and concentrate.

Twenty-six years later, Emerson recalled that June meeting in Chicago: "At that meeting, on motion of George W. Fuller, the famous Committee of One Hundred was authorized for the purpose of coordinating the efforts of interested groups and making definite recommendations for procedure. The Committee was appointed on July 15, 1927, and rendered its final report at a meeting of APHA in Cincinnati on October 21, 1927."

The membership list of the Committee of One Hundred is a roster of the Federation's heritage and reads like a "Who's Who" in engineering and science in the 1920's. (The complete list is included in the Appendix.) The Committee's recommendations were "(1) that a national sewage works association with an annual meeting would be undesirable at present, because travel expenses to an annual convention would be generally prohibitive; (2) that existing local sewage works associations should be strengthened and others formed in each state, or in groups of two or three states taken as a unit; (3) that the local associations should be joined into a nationwide federation guided by a central board of direction; and (4) that a journal should be published by the federation containing important papers presented at all the local meetings, together with proceedings of the meetings and original research articles, for distribution to every member of each local association at a moderate price." It is interesting to note that the four recommendations of the Committee of One Hundred differ only slightly from George W. Fuller's original concept.

According to Emerson's 1952 review, "This report fully endorsed the federation idea, indicated that research papers and operating data on plant costs and efficiency were available or in prospect to the extent of some 300 printed pages a year for the next few years, and that prospective advertising revenue and modest dues from members would be sufficient for printing and distribution costs."

The recommendations of the Committee of One Hundred were adopted at that Cincinnati APHA meeting and the Committee formally disbanded. In its place, Emerson was authorized to form a smaller committee to proceed actively on organizational details. This small committee met in January, 1928, in New York City at the annual meeting of the ASCE. It was more formally constituted on March 8 as the Implementing Committee, with four subcommittees under the chairmanship of H.W. Streeter of the Public Health Service, W.J. (Bill) Orchard of Wallace & Tiernan, John R. Downes of New Jersey, and Kenneth Allen of New York.

Among the Implementing Committee's 38 members, there were five New Jersey representatives, four New Yorkers, and five Iowans, with the remainder from other states. (The complete roster of the Implementing Committee is listed in the Appendix.)

ASCE Committee Report

The ASCE Eddy Committee also reported on the June 1927 Chicago meeting in its January 19, 1928, report. The report is almost a repetition of other committee reports noted above, but it is included here as a historical record.

"Following the Chicago Conference of June 11, Mr. C.A. Emerson, Jr., Chairman, appointed a Committee of One Hundred to 'survey present facilities and suggest improved program for discussing and publishing papers and data on treatment and disposal of sewage and industrial wastes and on allied subjects.'

> The report of this Committee [The Committee of One Hundred] was presented and its conclusions adopted at a conference held in Cincinnati, Ohio on October 21, 1927, at the close of the American Public Health Association Convention. In brief, the Committee found that:
>
> (1) There is need for an organization to publish papers and discussions on research, operating records, and abstracts.
>
> (2) Such an organization should take the form of a federation of State and sectional associations.
>
> (3) Existing associations will serve as a nucleus for such a federation.
>
> (4) Such local associations now exist in eight states and are under consideration in several others.
>
> (5) Adequate funds appear to be available for financing the publication of a suitable journal.
>
> A new committee was appointed to report upon the details of organization of a federation of local associations at a meeting to be held in New York, N.Y. on January 19, 1928, at the time of the Annual Meeting of the Society.
>
> After mature consideration and the interchange of individual views by correspondence, your Committee recommends the following policy and procedure for the Executive Committee:
>
> 1. The formation of a national society to hold meetings for the presentation of papers and discussions pertaining to stream pollution is not now justified and should be discouraged.
>
> 2. Moral support should be given to the movement for the formation of a federation of local associations now existing and to be formed in the future, for the purpose of publishing the results

of research, operating data, abstracts of articles published in other journals, and such other matter as will be of general interest to those working in this field.

3. Encouragement should be given to the presentation to the division of original papers comprising:

- (a) results of research which tend to establish fundamental principles of procedure in the solution of general problems.
- (b) analytical descriptions of permanent value of design and construction of works.
- (c) results of operation which demonstrate the practicability and efficiency of important processes and works and their sufficiency for the solution of the problems for which they were provided.

4. Papers should not be accepted by the Executive Committee which are confined to biological and chemical details of research and to ordinary operating data, as such papers are needed and can be handled appropriately by other organizations.

5. The Executive Committee should be prepared to appoint one or more members of a joint committee for the coordination of the work of the several organizations active in the field, if and when other organizations signify their desire for such a joint committee.

The purpose of the appointment of your Committee having been accomplished or being now carried forward by other agencies, a Committee on this subject need not be continued.

Respectfully Presented, Harrison P. Eddy, Chairman.

Emerson's historical review in 1952 reported the subsequent developments as follows: "Then came a new and smaller committee, which reported a definite program at a meeting in New York City on January 19, 1928, and in turn became an Implementing Committee appointed on March 8 and based on four sub-committees, as follows: Organization Committee: Harold W. Streeter, Chairman; Finance Committee: Wm. J. Orchard, Chairman; Publication Committee: John R. Downes, Chairman; Coordination Committee: Kenneth Allen, Chairman. A name was selected for the Federation and for the journal and all the Committees were put to work. Dr. F.W. Mohlman was selected as editor at a meeting of the Publication Committee in Trenton, N.J., on March 28 and in July, Howard E. Moses was appointed secretary-treasurer."

The Chemical Foundation

A milestone in the Federation's history was entry of the Chemical Foundation into its activities. The Foundation was a quasi-governmental agency formed during World War I to administer patents and processes seized by the Alien Enemy Property Custodian and to dispense the proceeds from them for furthering progress in science. The impounded royalties on the U.S. patent granted Karl Imhoff constituted some of this income and could be devoted to sanitation. Fortunately, Dr. Rudolfs was acquainted with W.W. Buffum, General Manager of the Foundation, and at a meeting of the American Chemical Society in St. Louis introduced him to a group of some five or six sanitarians gathered for lunch and steered the talk around to a discussion of the proposed Federation.

Buffum's interest in the Federation was aroused and on his return to New York arranged several meetings between Francis Garvin, president of the Chemical Foundation, and members of the Federation's New York group. These meetings resulted in a letter from Buffum on May 11, 1928, which stated that remaining funds available for sewage research could be spent to better advantage in furthering the Federation than in continuing the grants to individuals to finance specific research projects, as had been the Foundation's custom. Accordingly, the Chemical Foundation offered to take over the entire business management of the journal for a trial period of a year, underwrite any deficit, and turn over any profit. This added backing was promptly accepted, and the arrangement with the Chemical Foundation continued for more than 11 years, until Mr. Buffum's death in 1940.

First Board of Control Meeting

With the official beginning of the Federation nearly at hand, an organizational meeting of the Board of Control was held in the Stevens Hotel in Chicago, Ill., on October 16, 1928, and was attended by 11 representatives of eight associations that had voted to affiliate and had elected representatives. These eight associations were from Arizona, California, Central States, Iowa, Maryland, North Carolina, Pennsylvania, and Texas. (North Carolina had to reorganize before joining the Federation in 1929; see Chapter VII.

At this meeting, a constitution and bylaws were adopted, officers elected, and committees appointed.

The officers were: Chairman: C.A. Emerson
Vice-Chairman: Abel Wolman
Secretary-Treasurer: H.E. Moses
Business Manager: W.W. Buffum
Editor: F.W. Mohlman

The first issue of *Sewage Works Journal* appeared late in October 1928, and was sent to 273 paid subscribers.

Harris Seidel's research revealed that the first Board of Control faced the question of eligibility for affiliation of those associations that were both water and sewage groups, some heavily water-oriented. According to the Board minutes, "A short discussion took place concerning the status as to eligibility for membership in the Federation of a combined waterworks and sewage association. No formal action was taken, but it was the sense of the Board that the members of such combined associations primarily interested in sewage would form a separate group that would be eligible for affiliation with the Federation."

As a result, the North Carolina and Missouri Associations found it necessary to reorganize. Both accomplished this in November 1928, but were not listed as formally affiliated until 1929. By contrast, Texas remained in the 1928 "charter" group, although it did not carry through the suggested reorganization until January 1929.

New Jersey, which had contributed much to the development of the Federation, chose not to affiliate at the start, preferring to maintain its identity and independence as the oldest state association in the U.S. A conference group was formed in New Jersey to receive the journal, and this arrangement continued until 1942 when the New Jersey Association became formally affiliated with the Federation.

Several of the original founders of the Federation were from states that were not members of the Federation, for example, New Jersey, which had declined to join, and New York, which had not yet organized an association. To overcome the problem, the organization's by-laws were amended at the October, 1928, meeting to provide for five at-large directors. By this mechanism, Bill Orchard and John Downes of New Jersey were elected to the Board, along with W.W. Buffum and Kenneth Allen of New

York, and H.W. Streeter of the U.S. Public Health Service in Cincinnati.

Thus the Federation was organized after many months of discussion among members of the ASCE with contributions from APHA, CSSE, and AWWA. Mr. Emerson gave credit to George Warren Fuller for proposing the original concept. It is amusing and ironic that Fuller, who later became a "patron saint" of the American Water Works Association, did not prevail on AWWA to sponsor or participate in the development of the Federation. In 1927, it appears that AWWA wanted no part of wastewater.

History Committee Comment

Over the past several years, during the preparation of this history, many persons have commented that any history of the Federation must start with the Committee of One Hundred. As indicated by the material in this chapter, most of those 100 men did little or nothing but lend their names to the project. Although that was important, they were never involved in real organizing activity. As one of the contributors to this volume has said, "What in the world could any group of 100 men have done in a few months, scattered across the country as they were? There could have been some mail exchange, but no one has suggested that there was even much of that."

Undoubtedly, the man who thought up the Committee of One Hundred is the one who deserves credit for an extremely clever and fortuitous strategy. The concept of a Federation had already been developed based on George Warren Fuller's suggestions. The Committee of One Hundred was a way of getting national recognition for it, but it was the Implementing Committee that worked out the important details. As with many efforts, probably a small handful in that group of 38 men did most of the work. The authors of this history feel it is time to commend the work of the Implementing Committee.

It is the consensus of the Federation History Committee that most of the credit for forming the Federation should go to Charlie Emerson, with some strong assists from George W. Fuller for the concept and from Messrs. Orchard, Eddy, Moses, Streeter, Ehlers, Rudolfs, Downes, Mohlman, Allen, and others. It is appropriate that Federation Awards should be named for Emerson, Orchard, and Rudolfs.

First Elected Officers

Charles A. Emerson
Chairman

Abel Wolman
Vice Chairman

Floyd W. Mohlman
Editor

H. E. Moses
Secretary-Treasurer

Part of the assurance of income that would enable the *Journal* to meet expenses came from promises to purchase advertising. Pictured here are those included in the first volume as full page ads.

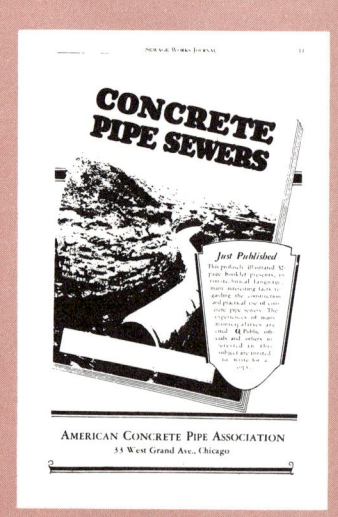

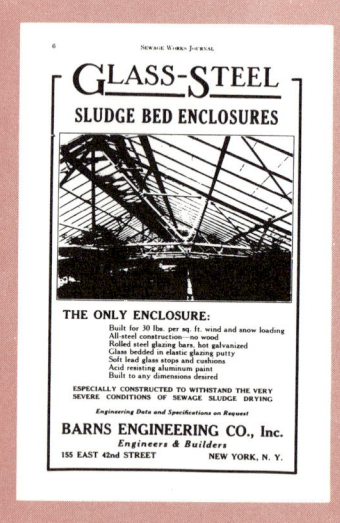

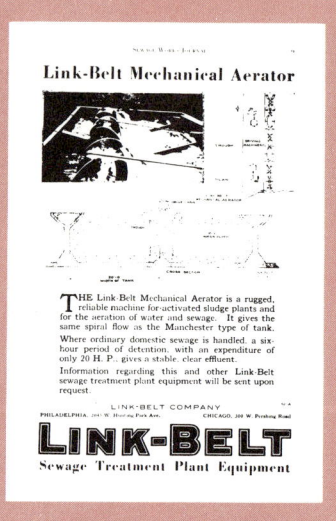

Master Construction Glass-Overs for Sludge Beds

The plant at Tenafly, N. J., using the activated system is a particularly interesting installation

35 Installations Indicate Our Experience Building Glass-Overs for Sludge Beds

THE facts are, that no one has built as many glass covers for sludge beds. The largest ones built, we built. That famed Boonton one, known as quite the last word, is the widest single house glass-over built.

It's a Lord & Burnham.

We don't claim to be sanitary engineers. Not a bit of it. But we do have men with engineering training who can sit down with you and talk your language. Men who specialize on sludge bed glass-overs. Men who have every advantage of our experience, acquired in the 35 installations made in less than 3 years.

These range from the smallest to the largest.

Such for instance, as Fostoria, Ohio; Greenwich, Conn.; Dayton, Ohio; Boonton, N. J.; Lackawanna, N. Y.

To a complete list you are welcome. Also there are binders giving special information on the Cleveland, Boonton and other installations. Send for it.

Lord & Burnham Co.
Sludge Bed Glass-Overs

Graybar Bldg., New York — 208 So. La Salle St., Chicago — Harbor Com. Bldg., Toronto

Branch Offices: Boston, Philadelphia, Irvington, Denver, Detroit, Cleveland, Buffalo, Greensboro, St. Louis, Montreal, St. Catharines

---

SEWAGE PLANTS

will receive better attention if the working parts are fully understood by the Plant Operators.

We publish seven (7) Catalogs on Air-Lock Apparatus for such plants. Write us for your copy.

SEWAGE DISPOSAL (SPRINKLER) PLANT

SOUTHERLY PLANT, CLEVELAND, OHIO

We furnished (3) Sets Twin Dosing Tanks 36" Siphons Deep Bell Type

Before designing a Sewage Plant or a Sewer System, Engineers should examine our catalogs showing apparatus

for

Sludge Pumping, or for Operating Bacteria Beds
Sand, Contact or Sprinkler type
Twin Dosing Siphons, Imhoff Tanks, also Miller Flush-Tank Siphons, Water Regulators, etc.

PACIFIC FLUSH-TANK CO.

9 Park Place, New York, N. Y. — 4241 Ravenswood Ave., Chicago, Ill.

HEALTHY CITIES *are* PROSPEROUS CITIES

SANITATION

SANITATION authorities are not content to rest with the purification of water supplies. They recognize the fact that there are other menaces to the health of cities. Untreated sewage is one. In this fight to effectively treat sewage and industrial wastes Liquid Chlorine is an important weapon. Some facts bearing on this important subject are discussed in E B G literature. Copies will be forwarded on request.

Look for the RED Cylinder

Your Water and Sewage Deserve the Best Treatment

Liquid Chlorine
Electro Bleaching Gas Co.
PIONEER MANUFACTURERS of LIQUID CHLORINE
Plant NIAGARA FALLS, N.Y.
Main office 9 East 41st Street New York

CHLORINE —*and* SEWAGE TREATMENT

W&T VACUUM SOLUTION FEED CHLORINATOR TYPE MSV

THE RESEARCHES of the past five years have demonstrated that chlorine has a wide field of usefulness in the treatment of sewage.

Scores of widely scattered plant scale installations have shown that chlorine—beyond being a definite germicide—effectively controls odors. Prechlorination of sewage will improve primary tank operation and will maintain the secondary treatment at high efficiency—at unexpectedly low costs.

Here as in all other fields of chlorination the merit of W&T *Solution Feed* Chlorinators has been demonstrated beyond question. May we send you a copy of "The Chlorination of Raw Sewage for Odor Control"?

"The only safe water is a sterilized water"

WALLACE & TIERNAN
COMPANY, INCORPORATED
Manufacturers of Chlorine Control Apparatus
NEWARK · NEW JERSEY

NEW YORK CHICAGO KNOXVILLE SAN FRANCISCO MINNEAPOLIS PITTSBURGH DALLAS KANSAS CITY LOS ANGELES SEATTLE ST. LOUIS BUFFALO HARRISBURGH INDIANAPOLIS DETROIT WALLACE & TIERNAN, LTD., TORONTO, CANADA WALLACE & TIERNAN, LTD., LONDON, ENGLAND

"METRO"

The Original Vitrified Filter Flooring

Since the Metropolitan Paving Brick Co. designed and introduced the first vitrified flooring for trickling filters, patented under the name "Metro," it has had unqualified success. Acceptance by Sanitary Engineers and Public Officials has been almost instantaneous. Modern, practical, economical "Metro" has proved all of these points under actual operating conditions.

It features thorough drainage and aeration with provision for flushing. Strength is ample for the load required. And from the cost angle, the first cost is the last. "Metro" flooring requires no attention.

PAVING BRICK COMMON BRICK FACE BRICK HOLLOW TILE FILTER FLOOR SYSTEMS

THE METROPOLITAN PAVING BRICK CO.
CANTON, OHIO, U. S. A.

CHLORINE

LIQUID CHLORINE
To meet your requirements

With the introduction of LIQUID CHLORINE, the Hooker Electrochemical Company came early to be a principal manufacturer of this product. As an aggressive leader in promoting its effective use [illegible]

Our new plant recently placed in operation at Tacoma, Washington, supplements our long established operation at Niagara Falls, where for nearly a quarter of a century we have been producing a long line of Industrial Chemicals of Highest Quality.

Our policy of carrying ample stocks of material and maintaining complete tank car, multi unit car and cylinder equipment at both of our plants, assures prompt and efficient service to all sections of the country.

The advice and experience of our technical and engineering staff are at your service. We solicit your inquiries and welcome the opportunity to assist you.

HOOKER PRODUCTS
Caustic Soda
Liquid Chlorine
Bleaching Powder
Muriatic Acid
Monochlorbenzene
Paradichlorbenzene
Benzoate of Soda
Benzoic Acid
Benzoyl Chloride
Benzyl Alcohol
Antimony Trichloride
Ferric Chloride
Sulphur Monochloride
Sulphur Dichloride
Sulphuryl Chloride
Salt

HOOKER ELECTROCHEMICAL COMPANY

Eastern Sales Office, 25 Pine St., New York City, Plant: Niagara Falls, N. Y.
Western Sales Office, Tacoma, Washington, Plant: Tacoma, Washington

---

Dorr Detritors in the Sewage Treatment Plant at Winston-Salem, N. C.

The Dorr Detritor
—the latest advance in sewage treatment practice

Consider these points:—

The Detritor continuously removes the grit from the sewage, washes it, and discharges it in a clean condition practically free from putrescible organics. The grit can be used on roads or as fill without nuisance.

For new plants the Detritor simplifies plant design, as varying velocities of the flow will not affect its operation.

The disagreeable manual task of removing grit from the channels is eliminated.

Positive, continuous removal means that the grit cannot interfere with subsequent steps in the process.

The Detritor is simple in operation and has no submerged bearings.

You should have a copy of our bulletin No. 6481, which describes the Detritor. Our nearest office will gladly send you one.

DORRCO

THE DORR COMPANY
ENGINEERS
247 PARK AVENUE NEW YORK CITY
INVESTIGATION TESTS DESIGN EQUIPMENT

DENVER, COLO. — CHICAGO, ILL. — LOS ANGELES, CAL. — WILKES-BARRE, PA. — JOPLIN, MO. — ATLANTA, GA.
TORONTO, ONT. — LONDON — BERLIN — PARIS — JOHANNESBURG, S. A.
MELBOURNE, AUSTRALIA

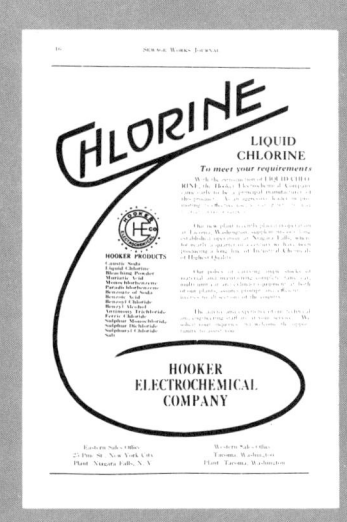

Karl Imhoff (center) on a visit to Atlanta, Ga., 1913, to inspect the Imhoff tank installation at the Peach Tree Creek plant. Imhoff patent revenues, administered by the Chemical Foundation, provided early financial support for the Federation of Sewage Works Associations. Shown with Dr. Imhoff are, William A. Hansell, engineer of sewers and Robert M. Clayton, city engineer of Atlanta.

It was George W. Fuller's early concept of what a national sewage works organization might be that prevailed and was implemented as the Federation of Sewage Works Associations in October 1928.

It was Bill Orchard, a member of the group that conceived the Federation, who obtained promises from 20 firms to advertise for one year in a sewerage journal. These promises along with support from the Chemical Foundation gave the assurance needed to begin publication of the *Sewage Works Journal*.

18

Chapter Two

The Emerson/Moses Era-- 1928-40

The Federation was formed primarily as an organization to provide a journal for the publication of papers on sewage works, and the first 12 or so years of its existence were devoted to little more than that objective. Its only meetings were those of the Board of Control, held in New York City every January in conjunction with the joint annual meetings of ASCE and the New York State Sewage Works Association.

Each Member Association had two Directors on the Board of Control, positions customarily filled by men who could conveniently get to New York City. The Directors' terms were not fixed, and some served many consecutive years without interruption.

Federation officers also continued generally to serve without interruption. Charles Emerson continued as chairman through the 1930's, along with H.E. (Ted) Moses as secretary-treasurer, W.W. Buffum as business manager, and Dr. Floyd Mohlman as editor. The only office to change hands was that of vice-chairman; the following men served the early Federation in that position:

Abel Wolman	1928 - 33	Morris M. Cohn	1936
Julius W. Bugbee	1934	Linn H. Enslow	1937 - 40
James R. Rumsey	1935		

In his luncheon speech at the Federation Silver Anniversary meeting in 1952, Emerson commented on the first 10 years of Federation activities as follows: "When the Federation was organized it was hoped to attain a membership of 1,000 in 10 years, but at the conclusion of the decade there was a net membership of 2,472 here and abroad, with 399 nonmember subscribers to the Journal, or nearly three times our rosiest expectations. The tenth anniversary was celebrated by publication of the 371-page volume entitled 'Modern Sewage Disposal,' which was produced through the untiring editorial efforts of that loyal member of the Central States Association, Langdon Pearse. Some 2,500 copies were sold, netting a welcome profit to the Federation."

Reorganization, 1940-41

By 1940, the Federation had grown to more than twice the membership its founders had anticipated. Although the Chemical Foundation was still providing complete business management for the *Journal*, the Federation had been carrying its own weight financially for several years. The burden of correspondence and committee work, however, was rapidly growing beyond what could reasonably be handled on a part-time voluntary basis. Much thought was being given to how the Federation could reorganize to best serve its growing role in the sewage works field.

In 1939, a blue-ribbon Committee for Expansion and Reorganization was appointed. This E&R committee presented its first report at the January 1940 Board meeting in New York, but was then asked to study certain items at greater length. The Committee's final report was completed in September. Proposed changes in the Federation organization and scope received thorough discussion in Chicago, first in an open business meeting and then in the Board of Control meeting which followed. Substantial agreement was reached, although there remained a number of details to be ironed out.

Emerson's 1952 historical review of the early years explained the transformation from a journal publisher to a full-fledged technical and professional organization: "The continued growth and increasing usefulness of the Federation during the first decade made it evident that the original type of organization, largely managed by part-time efforts of volunteer officers, could not hope to adequately meet the multitudinous needs of the workers in the rapidly growing and ever-changing sewage works field. Accordingly, a special committee was appointed and at the Annual Meeting on January 18, 1940, recommended a long-term program, which has since been adopted in full."

That meeting was followed closely by the 25th annual meeting of the New Jersey Sewage Works Association in Trenton, March 20-22, 1940. The New Jersey silver anniversary meeting was itself of national scope, with speakers from across the country. The attendance was a record-breaking 450, and 35 manufacturers exhibited equipment.

The interest in this New Jersey meeting was heightened considerably by W.J. (Bill) Orchard's prior announcement that he would be making a major policy statement urging a "strong, national sewage works association." This he did, proposing a more cohesive organizational structure (perhaps modeled after

AWWA) and a working relationship with the Equipment Manufacturer's Association, similar to that of AWWA.

The reaction was mixed. In the discussion following, some attendees were enthusiastic; others supported the concept of a strong national organization, but tempered their support. A typical comment was, "If this national association is going to take away the rights and powers of the local organizations, I am against it."

A resolution was offered, however, giving the New Jersey officers and executive committee the power to act either for or against the New Jersey Association joining in a new national organization. One program participant, expecting to see the national organization founded on the spot, was outspoken in his disappointment. The minutes recorded the following exchange:

> Mr. Levine: I came all the way from Iowa as representative of the Iowa Group understanding that this proposition of forming a National Association was to be taken up and considered . . . and I thought I would be able to tell the Iowa group that a national organization was to be a fact. As I interpret this resolution correctly, it leaves the matter exactly as it was.
>
> Mr. Orchard: The resolution refers the matter to the Officers and Executive Committee with power. This is a definite step of progress.
>
> Mr. Levine: I got the impression from the resolution that the matter stood as it was.
>
> Mr. Orchard: This is a meeting of the New Jersey Sewage Works Association. As a result of the discussion a resolution has been offered and is now on the floor to the members of the New Jersey Association giving the Officers and Executive Committee of this New Jersey Association power to affiliate with a national association without waiting for another Annual Meeting. This is a definite step in the right direction. Mr. Levine, you can tell your group in Iowa there will be a national meeting this Fall. . . .
>
> Without further discussion, the motion on the Resolution was put and the Resolution was unanimously adopted.

In his June, 1940, *Sewage Works Journal* editorial, Dr. Floyd Mohlman commented: "The growth of these various associations, each with its distinctive characteristics, has been the main feature of strength of the Federation and the plan has been too successful to warrant tampering with it."

A subsequent editorial in the *Sewage Works Journal* in September, 1940, brought into sharp focus the importance of the forthcoming Chicago convention and the issues facing the 12-year-old Federation. Dr. Mohlman wrote: "By far the majority of the members of the present Federation realize that a turning-point has been reached in its affairs. We can no longer stand still and refuse to accept the further responsibilities of time and money that are required to advance the prestige of the Federation. We need more activities, more cohesive strength on committee work, unity on national policies, a clearing house for sewage works information, aid for the poorly paid operators of small sewage works, publication of manuals and data on costs, more frequent publication of the *Journal,* and an annual convention that will bring together sewage works men from all over the country, in the one organization to which they should belong. Now we find them at conventions of the American Water Works Association, the American Public Health Association, the American Society of Civil Engineers, the American Chemical Society, the American Public Works Association, and others. In all of these the sewage men are only participants—why don't we draw our own family together in our own sewage works convention? Well, the answer is, we will, and the Chicago convention ought to develop only one regret—that we haven't had a national convention years ago."

Emerson said later, "Those who were present at the Hotel Sherman in Chicago in October, 1940, recall that attendance exceeded 600, that the 55 exhibitors were well pleased, and that the Convention was a success technically, socially, and financially. 1940 marked the start of activities of the Sewage Works Practice Committee, which to date under Morris M. Cohn has produced five valuable Manuals of Practice and has several others in course of production. It also saw Frank Woodbury Jones and the eight charter members of the Quarter Century Operators' Club seated in glory."

Late in the Chicago Board of Control meeting a special subcommittee was appointed consisting of five members of the former E&R committee. Their assignment was to report to the January, 1941, Board of Control meeting "with a plan and a report on ways and means to motivate and place the new Constitution in operation." The five men were Arthur S. Bedell, Chairman; Gordon M. Fair, F.W. Gilcreas, Max Levine; and W.J. Orchard, Secretary. This group soon became known as the Committee on Motivation.

This small committee was faced with some real challenges. Among them were a new dues structure; affiliate (non-Federation) memberships within Member Associations; and the question of how the Federation could retain and encourage membership of low-income operators in the Federation itself. All these questions but the last were resolved and accepted by the January, 1941, Board of Control. The question of membership for low-income or small plant operators was discussed at length and then finally referred to the Executive Committee for further study, with instructions to submit a proposal to the Board of Control for a letter ballot later that year. The result was the "Alternate Active" member classification under which an operator could receive every other issue of the *Journal* for dues of $1.50 per year.

Three months later, on January 15, 1941, at a special meeting of the Board of Control, authorization was given for establishment of an Executive Secretary's office, with William H. Wisely to serve as Secretary on a half-time basis at the start, and for establishment of closer cooperation with the Water and Sewage Works Manufacturers' Association.

At this meeting, the yeoman work of Howard E. Moses, who had served as Secretary since the inception of the Federation, was recognized by his designation as Honorary Secretary.

The new constitution and by-laws were also adopted on January 15, 1941. This was the last January Board meeting; subsequent meetings of the Board of Control were held in the fall during annual Federation conferences.

The Committee on Motivation, through its Secretary, W.J. Orchard, had very substantial influence on the shape and direction of the reorganized Federation. In addition to bringing in ten specific recommendations on such matters as structure, dues, and the establishment of a headquarters office, this group actually served as the first nominating committee. Following acceptance of its report, the committee proposed a complete slate of officers for terms extending to October, 1941. Next, the committee was ready with a suggested roster of members for all the new constitutional committees.

The changes were far-reaching. Active member dues were doubled from $1.50 to $3.00, effective in 1942. Officers were limited to one term, and were not eligible for re-election. Emerson was elected President for the 1941 (January-October) term, and Arthur Sidney Bedell of the New York State Health Department was elected Vice-President. In October 1941, Bedell was elected President and George J. Schroepfer of Minneapolis was elected Vice-President.

The new constitution also provided that each Member Association would have only one Director on the Board of Control instead of two. At the January 1941 Board meeting, the terms of the New Directors were distributed by lot evenly between one-, two-, and three-year terms, all beginning in January, 1941.

Another important step was the decision to begin loosening the ties with the Chemical Foundation, which had served the Federation long and well.

William W. Buffum, the Federation's Business Manager from its inception, had died on June 22, 1940. By letter ballot the Board of Control adopted an appropriate resolution, copies of which were sent to Buffum's family and to the Chemical Foundation, and published in the July, 1940 issue of *Sewage Works Journal*. Arthur A. Clay, who succeeded Buffum as Treasurer and General Manager of The Chemical Foundation, was designated by the Foundation as the new Business Manager of the Federation and, by letter ballot, was elected to fill the unexpired term of Buffum as Member-at-Large on the Board of Control.

Ted Moses' Last Report

At the January 15, 1941, Board meeting, H.E. Moses gave his last official report as Secretary-Treasurer of the Federation. He had served 12 years and 3 months in that capacity. Moses' report includes tabulations of gross and dual memberships and records of dues paid. It did not, however, cover the year-by-year growth of Federation membership; the data in Table II-1 were developed to establish that aspect of the Federation's history.

There were a number of difficulties in trying to put together an accurate tabulation of Federation membership during the early years. Perhaps statistics had a lower priority then. The first record lists a total membership of 411 for the seven "charter" associations in the fall of 1928. Beginning in 1930, annual Directory issues of the *Journal* listed names of all Member Association members; starting in 1933, addresses were also given. These lists provide the only detailed record of membership until the publication of a tabulation in the March 1936 *Journal*.

For some years these lists included all dual members—those who held memberships in one or several other Member Associations besides their "home" associations. It is almost certain that early membership totals as listed include these duplications. The individual association membership figures shown for 1934 and subsequent years are net, after deduction of dual members. Using the best available evidence, an attempt was made to estimate net active Federation member totals for the years 1929-32, as shown in Table II-1.

These membership data are for the end of each calendar year. The original Maryland Association was expanded into the Maryland-Delaware Association in 1930. A New Jersey Conference Group was organized in 1929 to receive the Federation *Journal*; the Conference Group was disbanded when the New Jersey Association affiliated with the Federation in 1942. The Dakota Association was admitted by the 1936 Board of Control with each of the states comprising a section; in 1960 a separate Member Association was formed in each state and affiliated with the Federation.

In the early years there were hundreds of nonmember subscribers to the *Journal* in addition to members included in these tabulations. These subscribers included libraries, companies, municipalities, and individuals both in this country and abroad. The number of nonmember subscribers grew as the Federation membership grew.

The Federation had no office of its own prior to 1941. During those years, the secretarial records were maintained in Moses' office in Harrisburg, Pa.; editorial work was handled in Mohlman's office in Chicago; the business of the *Journal* continued to be managed in the Chemical Foundation's office in New York

City; and the *Journal* was printed in Lancaster, Pa. This required a great deal of coordination and long-distance management.

Emerson's accomplishments have been honored by the creation of the Charles Alva Emerson Medal for outstanding service to the Federation. Moses was Assistant, and later Chief, Engineer of the Pennsylvania State Department of Health. When the Federation was reorganized in 1941, he was made Honorary Secretary. Both Emerson and Moses contributed greatly to Federation growth for the rest of their lives.

Emerson died at the age of 73 in August 1955. For the last 15 years of his life, he was a senior partner in the consulting firm of Havens and Emerson. A brief biography is of interest because his life spanned an important period in environmental engineering.

Emerson was born in Beloit, Wisconsin, on July 10, 1882. He attended Beloit College, and in 1903 he was graduated with a B.S. degree *cum laude*. Thereafter he attended the Massachusetts Institute of Technology, where he received a B.S. degree in Sanitary Engineering in 1905. Later, he took a graduate course in public health administration at Johns Hopkins University, while serving with the Pennsylvania State Department of Health as principal assistant engineer.

Emerson began his career as assistant engineer on the design of the Columbus, Ohio, softening and filtration plan in 1905. He then joined the Baltimore Sewerage Commission in 1906, where the largest wastewater treatment plant in the country at that time was to be built. In 1911 he moved to the Pennsylvania State Department of Health and 3 years later (1913) became its chief engineer, a position he held for 9 years. In 1923 he joined the firm of Fuller & McClintock, internationally known consulting sanitary engineers. After the death of George W. Fuller, Emerson became associated in 1936 with the Cleveland firm, George Gascoigne and Associates. After the death of Gascoigne, Emerson in 1940 entered into partnership with William Havens to form Havens and Emerson, consulting sanitary engineers with offices in Cleveland and New York.

During his professional career Mr. Emerson served on numerous technical committees, authored many papers, and was

TABLE II-1 FEDERATION GROWTH
EARLY MEMBER ASSOCIATIONS, 1928 - 1940

Member Associations	1928	1929	1930	1931	1932	1933	1934	1935	1936	1937	1938	1939	1940
Arizona	15	15	13	6	9	16	16	16	8	22	26	26	28
California	120	207	225	237	221	219	178	175	220	233	251	263	279
Canada						27	42	62	66	80	119	121	121
Central States	62	105	158	150	150	137	156	177	225	344	420	503	497
Dakota-North										26	12	16	7
Dakota-South										22	24	28	23
Federal	57		22	29	29	27	31	47	52	61	54	60	63
Georgia									45	37	43	34	43
Iowa		40	32	40	39	28	25	35	39	32	27	35	21
Kansas								23	30	26	28	25	24
Maryland-Delaware	44	50	36	46	43	42	33	28	19	20	23	28	27
Michigan			33	53	67	65	77	84	89	100	97	147	136
Missouri		84	100	103	95	65	71	53	90	37	28	18	10
New England		96	101	102	140	136	129	132	140	139	143	159	175
New Jersey Conference		48	44	47	41	55	52	56	57	58	63	61	68
New York		186	218	237	248	259	261	306	406	443	518	558	608
North Carolina		52	104	104	84	73	77	74	57	85	55	71	79
Ohio					61	78	86	92	99	100	106	110	106
Oklahoma		21	13	13	9	10	4	2	1		0	4	1
Pacific Northwest								32	36	37	48	64	80
Pennsylvania	85	126	118	127	126	116	127	144	156	168	178	193	204
Rocky Mountain									7	28	28	35	38
Texas	28	77	56	35	35	27	14	11	10	40	24	25	28
UK-Inst. San. Engrs.					20	30	42	51	54	59	51	50	54
UK-Inst. Sew. Purif'n.					8	44	47	67	91	105	105	109	97
Argentina									33	24	1	12	2
Net Total, dual memberships deducted	411	1080	1220	1300	1400	1412	1468	1667	2030	2327	2472	2755	2819
No. of Member Associations	7	13	15	15	18	19	19	21	24	25	25	25	25

Notes: 1928 data taken from tabulation made in that year; later data from Directory counts. 1923-33 data include dual members (more than one Association), and thus totals are approximate; later data exclude dual members.

the recipient of many honors. An honorary Civil Engineer degree was awarded to him by Pennsylvania College in 1917.

In 1928, Emerson took over the work of George W. Fuller in organizing the Federation of Sewage Works Associations and for 11 years served as chairman of the Board of Control. When the Federation began holding annual meetings, Emerson was elected its first president; one of the Federation's principal awards, the Charles Alva Emerson Award, was established in his honor. He had continued to serve the Federation in numerous capacities, the chief of which was chairman of the Constitution and Bylaws Committee. In recognition of Charlie Emerson's service to the Federation he was named its first Honorary Member.

Emerson was a member of a number of technical and professional associations and had served as contributor to such texts and reference works as "Sewage Disposal," "Solving Sewage Treatment Problems," and the "Manual of Water Works Practice."

Chapter Three

The Coming of Age--1941-54

After the Federation's first 12 years, the Chemical Foundation no longer provided the management and financial aid that had carried the enterprise through its early years. On March 1, 1941, the secretariat in Harrisburg, Pennsylvania, and the business office at 654 West Madison Avenue in New York City were consolidated in the offices of the Urbana-Champaign Sanitary District on the outskirts of Urbana, Illinois. Relocation was a simple matter; the physical assets comprised only two transfer cases of files and a check for something less than $2,000. Through the courtesy of the trustees of the Urbana-Champaign, Ill., Sanitary District, the next Executive Secretary, William H. Wisely, was authorized to devote half of his time as Engineer-Manager of the District to the service of the Federation. The new (rent-free) office was established in the administration building of the sewage treatment works of the Sanitary District.

One tie with the Chemical Foundation was retained, however. Arthur A. Clay, Business Manager of the Foundation, agreed to continue handling advertising production for the *Sewage Works Journal* in return for an annual fee of $750. Dr. F.W. Mohlman in Chicago continued as editor of the *Journal,* a post he had filled with distinction from the *Journal's* beginning. This rather dispersed arrangement became effective in March 1941.

In prior years, the primary objective of the Federation had been to produce a viable technical publication, one sorely needed at the time to further the wide range of science and technology applicable to the prevention and abatement of water pollution. The technical excellence of *Sewage Works Journal* had surpassed all expectations, but its financial base was tenuous at best. There were no resources to support other membership and public services envisioned by the Committee on Expansion and Reorganization in 1940.

The vision of the E&R committee was largely fulfilled in the next 14 years with the achievement of fiscal independence and stability; distribution of the *Journal* on a monthly basis; broadening of the spectrum of publications; initiation of a modest offering of headquarters service; expansion of technical and professional committee activity; a program of Member Association visitation; and acceptance of responsibility in representing public interest in legislation and government regulations pertaining to water pollution management. This period, from 1941 to 1955, saw the important consolidation and strengthening of the Federation. It was a dynamic, vigorous, and stimulating time.

General Development

The 1941 reorganization was propitious, even though it was effected only 9 months before the attack on Pearl Harbor. Despite national concern with the entry of the U.S. into World War II, the status of the Federation gradually began to improve. Wisely, Mohlman, and Clay worked smoothly and effectively together, despite their geographical separation.

The Federation was legally incorporated under Illinois law in 1941, and its exemption from federal income taxes was promptly established under Section 501(c)(3) of the Internal Revenue Code, a designation that covers nonprofit educational, scientific, and charitable organizations.

From 1941-43, the headquarters staff consisted only of a half-time Secretary, one stenographer-clerk, and a part-time stenographer. By 1943, however, the growing membership, correspondence, and administrative detail had increased the demands on the secretariat to such an extent that they could not be satisfied by this arrangement. Net worth of the organization had risen from $3,000 to $22,000 in 3 years. The time had come for the Federation to take its final step toward independence.

At the annual meeting of the Board of Control in 1943, it was decided that Executive Secretary Wisely would assume that office on a full-time basis, and at the same time relieve Dr. Mohlman as Editor of the *Journal*. The experience and guidance of Dr. Mohlman was retained by naming him Advisory Editor. Advertising production services being furnished by Arthur Clay and the Chemical Foundation were also taken over in the Federation's first true headquarters office. The rented accommodations comprised four tiny rooms in the Illinois Building, located in downtown Champaign, Illinois. The transition was completed on January 1, 1944. Two more staff members were authorized, and the Executive Secretary was now assisted by an editorial assistant, a bookkeeper, and a stenographer-clerk.

No structural change in the organization of the Federation occurred through the next decade, but a significant action in 1950 clarified its objectives. Although the name "Federation of Sewage Works Associations" implied concern only with the waterborne wastes carried by public sewer systems, the interests of the membership and the content of the *Journal* encompassed waterborne industrial wastes as well. The proposal in 1950 to change the name of the organization led to vigorous discussion of the several choices; the name "Federation of Sewage and Industrial Wastes Associations" was finally adopted. It is noteworthy that the name "Water Pollution Control Federation," which was to be adopted in 1960, was suggested and strongly supported by the Executive Secretary-Editor at the time. It was rejected in 1950 on the premise that it sounded too much like a federal regulatory agency.

By the end of 1954, Federation membership had more than doubled and the annual operating budget had increased eightfold over that at the beginning of 1941. The headquarters staff then numbered six full-time persons in addition to the Executive Secretary-Editor: Executive Assistant, Associate Editor, Editorial Assistant, accountant, and two stenographers. Clearly, the staff had little spare time, and the pressures for service continued to grow.

TABLE III-1. Federation Growth, 1940 - 54

Year	Number of Member Associations	Number of Members	Annual Budget	Net Worth	Staff
1940	25	2 820		$ 3 076	
1941	26	2 850		$ 7 090	3
1942	26	2 391	$ 20 150	$13 490	
1943	26	2 544	$ 21 800	$21 982	
1944	27	3 034	$ 28 050	$28 300	5
1945	27	3 237	$ 29 625	$33 779	
1946	29	3 703	$ 36 300	$31 433	
1947	33	4 367	$ 38 700		
1948	34	4 903	$ 45 850		
1949	36	5 218		$44 298	
1950	37	5 138		$54 971	6
1951	37	5 139	$ 77 000	$64 444	
1952	36	5 450	$ 81 600	$67 530	
1953	37	5 762	$ 84 900	$71 909	
1954	38	6 346	$100 300	$75 802	7

Growing Statistics

The growth of an organization is not measured solely by the number of members it has or how much money it spends, but rather by what it does and the services it provides. Nevertheless, a statistical summary of the Federation membership growth is of interest, and the data in Table III-1 say much about this period in Federation history.

Committee Functions

The original constitution of the Federation provided for five basic committees: Executive, headed by Chairman Emerson; Membership, chaired by Harold W. Streeter (who drafted the original constitution); Publications, with Dr. Mohlman as chairman; Coordination, with Kenneth Allen as chairman; and Research, chaired by Max Levine. A revision of the constitution in 1937 reduced this group to the Executive, Publication, and Membership Committees, with the Research and Finance Committees designated as special committees.

Other early committees were the Committee on Standard Methods of Sewage Analysis, dating from 1931 when it was chaired by E.J. Theriault, and the Committee on Sewage Works Nomenclature, headed by Frank Woodbury Jones. The latter committee worked with similar bodies in ASCE and AWWA.

Professor Earle B. Phelps became chairman of the Research Committee in 1931, and the next year he produced the first annual "Critical Review of the Literature on Sewage Chemistry and Sewage Treatment." This valuable source was developed further by Dr. Willem Rudolfs when he became chairman of the Research Committee in 1938, and by his successor, Dr. H. Heukelekian. The annual literature review continues to be a *Journal* feature.

The 1941 expansion movement was manifested to a considerable extent by increased committee activity, both administrative and technical. The constitutional committees were as follows:

Executive: President Arthur S. Bedell, Chairman
General Policy: C.A. Emerson, Chairman
Publication: F.W. Gilcreas, Chairman

Organization: G.R. Frith, Chairman
Sewage Works Practice: Morris M. Cohn, Chairman
Research: Dr. Willem Rudolfs, Chairman

The Sewage Works Practice Committee was a significant addition, as this body directed the manual of practice program that began soon thereafter.

Other special Committees authorized by the Board in 1941 included the following:

Administrative
Finance Advisory Committee
Convention Management Committee
Convention Meeting Place Committee
Awards Committee

Technical
Operators Qualifications Committee
Operation Report Committee
Committee on Nomenclature
Standard Methods Committee

Professional
Publicity and Public Relations Committee
Civilian Defense Committee
Postwar Planning Committee

It is worth noting here that for the first time committee activity of a professional nature, directed to public interest and service, was being undertaken. The wartime committees (Civilian Defense and Postwar Planning) were both productive, but the public relations committee activity was suspended in 1944 and its responsibilities assigned to the staff.

More new committees followed from 1942 through 1947:
Committee on Honorary Members (1942)
Industrial Wastes Committee (1943)
Legislative Analysis Committee (1945)
Nominating Committee (1946)
Constitution and Bylaws Committee (1947)

Joint Committees

The desire of the Federation to collaborate with other responsible organizations in accomplishing mutual goals was demonstrated by its work with AWWA and APHA on "Standard Methods" and its cooperation with ASCE and AWWA in standardizing sewage works nomenclature. Both of these ventures began in the early 1930's.

Indicative of the developing stature of the Federation was its acceptance in 1947 of representation on the APHA Policy Advisory Committee. This was followed in 1950 by participation with AWWA, APHA, and WSWMA in the Water, Sewage, and Industrial Joint Committee on Critical Materials. In 1951 came the Joint Committee on Chlorine Supply, which included representatives of AWWA, the Conference of State Sanitary Engineers, and the Federation.

The Joint Committee on Advancement of Sanitary Engineering, formed in 1952, was comprised of members from ASCE, AWWA, APHA, the American Society for Engineer Education, and the Federation. The deliberations of this body resulted in the organization in 1958 of the first professional certification board in engineering, the American Intersociety Sanitary Engineering Board and its American Academy of Sanitary Engineers. In 1973, the two combined to form the American Academy of Environmental Engineers.

Authorization and appointment of committees does not by any means ensure that an organization will be more productive. Any committee is only as effective as its leadership and personnel. Almost all of the committees created at this stage of the Federation's development were successful in fulfilling their missions, and most of them are still active in the 1970's, though in some cases under different names.

Service to Members and the Public

The Federation always considered itself a technical organization rather than a professional society. Its membership, however, always included many engineers and scientists of professional stature. Furthermore, the management and operation of water pollution control facilities involves a dedication to the public welfare that is highly professional in nature.

There was an early recognition of the need for encouragement of professional attitudes within the membership, and of the need for greater appreciation and understanding of the public toward water pollution control. Such recognition led to the Federation's emphasis on improvement of sewage works operation, but because of the Great Depression and the prospect of World War II, the public was apathetic toward funding the construction and operation of pollution abatement works.

Membership Services

A well-intentioned, although somewhat premature, effort toward career development activity on behalf of members was the employment service initiated as a feature of the *Journal* in 1939. The secretariat attempted to serve as a clearing-house for both employers and job-seekers, and the effort was modestly successful. The Federation did not undertake any other activity directed toward the economic welfare of its members, but useful studies of plant operator salaries were made by the New England Sewage Works Association in 1947, the California Sewage Works Association in 1948, and the Iowa Sewage Works Association in 1955.

The Committee on Operator Qualifications produced several valuable reports on the education and training requirements for operators of various types and sizes of sewage treatment plants. These guidelines in the period 1943-48 were beneficial to state agencies engaged in licensing and certification programs.

After 1941, an effort was made to encourage participation by the officers and the Executive Secretary in the meetings of the Member Associations, thus bringing the Federation closer to the individual members. These meetings provided opportunities to build esprit de corps and to provide inspiration and encouragement to plant operators and managers. The objectives were also furthered through editorials and papers published in the *Journal*.

The new full-time secretariat in 1944 was ambitious despite its small staff and resources. An editorial entitled "At Your Service," published in the January 1944 *Journal* invited readers to call the new office for any technical and professional service it could render. This offer reflected the basic policy of the headquarters staff to give high priority to the conscientious handling of service inquiries.

Public Service

The 1941 reorganization of the Federation was implemented

a few months before the entrance of the U.S. into World War II. It is not surprising, therefore, that the earliest public affairs action involved the Committee on Civilian Defense and the Committee on Postwar Planning. Both of these efforts were pursued energetically and successfully. A 1943 report by the Committee on Civilian Defense provided sound guidance to public administrators concerned about possible sabotage of public services.

In 1942, the War Production Board requested the Federation to supply estimates of materials required for maintenance and operation of public sewage works. A survey made under the direction of George J. Schroepfer (Federation President, 1942-43) produced the desired information. There were other opportunities between 1942 and 1945 to cooperate with the WPB, the War Manpower Commission, and the Office of Civilian Defense.

Perhaps the first action of the Federation that reflected its national stature was a resolution adopted on October 24, 1942, which urged the Selective Service Board and War Manpower Commission to include key personnel in municipal sewage works departments in the list of critical occupations exempt from military service.

The Legislative Analysis Committee, appointed in 1945 "to study legislation pertinent to the interests of the Federation" and to advise the Board of Control on such legislation, was a major breakthrough. This action established the determination of the Federation to concern itself with legislation in the public interest. The new committee lost no time in considering the endorsement of the federal anti-pollution legislation designated as the Barkley-Spence Bill. The principle of primary responsibility at the state level supported by federal authority was endorsed in both the 78th and 79th Congresses. Review of all national and state legislation dealing with water pollution continued to be a primary function of this committee.

Material shortages caused by the Korean War in 1950 again led to action by the Federation to ensure needed allocations of materials for sewage works construction, operation, and maintenance. In that year the Board of Control adopted a resolution urging Congress, the National Security Resources Board, and the National Production Authority to establish such priority allocations.

Public interest activity of the Federation in this period was not limited to legislation. One venture in the early 1950's was an effort to bring about voluntary action on the part of synthetic detergent manufacturers to reduce the foaming effect of these materials in all sewage treatment processes requiring aeration. A letter from the Executive Secretary to the leading detergent manufacturers urged them to cooperate in solving the problem before regulatory or legislative restrictions might be imposed upon them. The response ranged from indignation at the thought that their products affected sewage works to a blunt "mind your own business" attitude. However, the Federation's foresight in this issue was demonstrated in the 1960's, when excessively foaming detergents became a major water pollution issue.

To be sure, the membership and public service functions of the Federation did not have great impact or results during the period 1941-55. The period was significant, however, because it brought the need for such services into focus, and initiated the attitudes and policies that gave the organization a strong professional dimension that continues to exist.

Membership Development

As indicated in the table on growth statistics, at the end of 1940 the Federation comprised 25 state and regional Member Associations having an aggregate membership of 2,820. Each Member Association established its own rate of annual dues, usually only nominally higher than the $1.50 per member per year that was payable to the Federation to cover the *Journal* subscription and all other expenses at the national level.

The 1941 expansion program was accompanied by an increase in the Federation's assessment to $3.00 per member per year, effective in 1942. There was grave concern that doubling of the rate would have serious impact on membership growth. Happily, this adverse impact did not materialize. While the growth curve flattened slightly for about 3 years, it rose at an unprecedented rate after 1945. It is unlikely that any other technical organization offered its members such a bargain as did the Federation at this time.

The 23 U.S. Member Associations in 1941 included 34 of the states; in addition there were three foreign Member Associations, two in England and one in Canada. Membership promotion efforts encompassed both the admission of new Member Associations and the increase of membership in the associations already affilated. The *Journal* was the main inducement to new Member Associations, as the non-member subscription rate was $5.00 compared to the annual assessment per member of only $3.00. In the 14-year period following the 1941 reorganization, the number of Member Associations grew to 38—32 U.S. associations covering 46 states and the Commonwealth of Puerto Rico, and six foreign associations located in England (which had two), Canada, Germany, Sweden, and Switzerland.

Individual memberships were developed through the Member Association secretaries and by taking measures to reduce losses through resignation and non-renewal. Good working relationships with the association secretaries were furthered through personal contacts and manuals providing guidance in membership development. Visits to Member Association meetings by the Executive Secretary were important in this process.

It was not until 1954 that a personal working rapport was established with the five European Member Associations. In that year the Executive Secretary participated in a meeting of the Institute of Sewage Purification at Blackpool, England, and made official contacts with other Member Associations in England, Germany, and Switzerland. During this trip honorary membership certificates were presented formally to William T. Lockett and Dr. Karl Imhoff, and many research institutions, government agencies, and sewage works were visited.

The position to which the Federation had risen by this time is evidenced by the account of that meeting in Blackpool published in the July-August 1954 issue of *The Water and Sanitary Engineer:*

> Immediately following the Presidential Address came one of the highlights of the whole Conference proceedings, and one which we enjoyed perhaps more than any other contribution. This was the Address by Mr. W.H. Wisely, secretary of the Federation of Sewage and Industrial Wastes Associations. A more polished or sincere contribution to an Institute Conference was never made and in his Address Mr. Wisely not only made his official introduction to members of the Institute of Sewage Purification and delegates from overseas, but more important, made an introduction

that caused him to be warmly taken up by every single member of the Conference.

Mr. Wisely was exceptionally well received not only officially in Conference but socially on the floor of the Hall during the recess. Mr. Wisely has that particularly charming facet of character of being essentially a friendly man, and that with members at all levels in the Institute. He conversed with and made friends with many delegates and was whole-heartedly accepted as one of them, no mean tribute to a newcomer to ISP Conferences! Without any doubt his visit to Blackpool was an unqualified success and we can only hope that he and men like him will continue to come to these meetings and refresh us with their personalities and knowledge of the art.

At the conclusion of Mr. Wisely's Address, very ably acknowledged on behalf of the Institute by Mr. Kershaw, Mr. Wisely presented to Mr. W.T. Lockett the Certificate of Honorary Membership of the Federation bestowed by the Board of Control, for his distinguished services in the sewage treatment field.

Honors and Awards

Recognition of meritorious accomplishment is cherished by everyone who takes pride in the quality of his work. Professional people, in particular, take pride in honors conferred by their peers. The establishment of a program of honors and awards was, therefore, a logical consequence of the Federation's move to a broader operational base.

The first step in this direction was creation of the grade of Honorary Member in 1941, to recognize eminence in a field within the stated objectives of the Federation. Appropriately, Charles A. Emerson was the first Honorary Member elected. This honor is a high one, as demonstrated by the fact that only 15 Honorary Members were named up to 1955; a limit was set of no more than three persons to be so honored in any year thereafter.

Other awards were soon established:

Charles Alvin Emerson Medal (1943), for service to the Federation.

Harrison Prescott Eddy Medal (1943), for outstanding research.

George Bradley Gascoigne Medal (1943), for solution of difficult operation problems.

Kenneth Allen Award (1943-48), for service to a Member Association.*

Arthur S. Bedell Award (after 1948), for service to a Member Association.

William D. Hatfield Award (1946), for performance in sewage works operation.

Willem Rudolfs Medal (1950), for a contribution to industrial waste control.

In a span of less than 10 years a sound program of awards was developed to recognize noteworthy accomplishment in most of the major areas of Federation interest. The program was successful in its two-fold purpose: to reward the initiative, ingenuity, and dedication of today's leaders, and to inspire even greater achievement by tomorrow's. In subsequent years, additional awards were created by the Board for specific outstanding services by members. (See Chapters IV and V and the Appendix).

In Retrospect

By 1954 the Federation was serving 6,346 members, an increase of 120 percent after the dues assessment was doubled in 1942; associate members totaled 68. This was considered quite satisfactory at the time in view of the early state of water pollution control as a technical specialty.

From its 13th to its 27th year, the Federation came of age in every respect. Its growth was dynamic; its technical publications, meetings, and committee services matured; and its fiscal integrity was established. In addition, the process of professionalization was begun and carried to an effective level.

Toward the end of the period, the Federation increased its interest in fundamental research. This interest was evidenced at the Board of Control meeting on October 9, 1952, when the Board entertained a proposal that the Federation give greater emphasis to the encouragement and stimulation of research in its areas of interest. Toward that end, in January, 1953, the Ex-

---

* The Award was renamed the Arthur S. Bedell award in 1948, in honor of the Federation's second President, because the NYWPCA had a Kenneth Allen award predating the Federation Award.

ecutive Secretary prepared a "Special Report on Research Activity" to provide a basic analysis that would lead to further consideration of the proposal. (The complete text of that Special Report appears in the Appendix.)

As this period ended, it could not be foreseen that only 5 years later an unprecedented wave of public concern for environmental quality would begin to engulf the nation. Water pollution would not be spurned by the politician but would be embraced by him as a popular cause. Enormous federal appropriations with equally awesome superstructures of bureaucratic regulation were to be forthcoming.

Fortunately, the Federation was firmly founded in its organization, policies, procedures, and traditions when the environmental concern became a national issue. The water pollution abatement movement was accelerating in momentum, but it sorely needed sound direction and rational perspective in its relationship to other pressing national problems. The Federation was ready to provide that source of perspective.

The Federation's second Secretary, William H. Wisely, resigned at the end of 1954 to become Executive Secretary of the American Society of Civil Engineers. Thus ended the "period of maturation."

Federation Presidents and Executive Secretary 1941-1954

On adoption of a revised Constitution and Bylaws January 15, 1941, the Federation elected its first President and Vice President who were not allowed to succeed themselves. Their terms of office started immediately and lasted until their successors were elected in October 1941. During this period the election of vice presidents to the presidency was practiced.

C. A. Emerson
President
1941

A. S. Bedell
President
1941-42

G. J. Schroepfer
President
1942-1943

A M Rawn
President
1943-44

A. E. Berry
President
1944-45

J. K. Hoskins
President
1945-46

45

F. S. Friel
President
1946-47

G. S. Russell
President
1947-48

V. M. Ehlers
President
1948-49

A. H. Niles
President
1949-50

R. E. Fuhrman
President
1950-51

E. Boyce
President
1951-52

E. S. Chase
President
1952-53

L. J. Fontenelli
President
1953-54

W. H. Wisely
Executive Secretary
1941-54

Charles Alvin Emerson Medal

(1943)

awarded for outstanding service to the Federation

Harrison Prescott Eddy Medal

(1943)

awarded for outstanding research

George Bradley Gascoigne Medal

(1943)

awarded for solution to difficult operations problems

Willem Rudolfs Medal

(1950)

for a contribution to industrial waste control

Kenneth Allen Award

(1943-48)

for distinguished service to a member association

William D. Hatfield Award

(1946)

for outstanding performance in sewage works operation

Arthur S. Bedell Award

(Since 1948)

for distinguished service to a member association

Chapter Four

Headquarters Move to Washington-- 1955-69

Wisely's successor as Executive Secretary-Editor was Dr. Ralph Fuhrman, who had been President of the Federation in 1950-51. Fuhrman was Superintendent of Water Pollution Control and Deputy Director of Sanitary Engineering for the District of Columbia. The selection committee consisted of Bill Orchard, long-time Chairman of the Finance Committee, as Chairman, Charles Emerson, the first Federation President, and Lou Fontenelli, the immediate Past President.

The loss of Pete Wisely as Federation Secretary-Editor at the close of 1954 presented many problems for the Federation. As its first full-time Secretary in charge of a staff that had grown to seven by the end of his term, he had dealt with many new situations and established working procedures. During those early years, he could supervise the work of each person on the staff and know in detail what each was doing as the Federation grew and developed. With Dr. Floyd Mohlman as advisory editor, a high level of technical and editorial competence was reflected in the Federation journal, then titled *Sewage and Industrial Wastes.*

Indeed, the *Journal* was the primary product of the Federation's efforts. Nearly all the staff's time was spent on its editorial production, advertising, and distribution. By the beginning of 1955, the total subscriber list for the *Journal* was 7448 (6346 members, 68 associates, and 1034 nonmember subscribers).

Dr. Fuhrman began working with the Federation at its office in Champaign, Illinois on December 1, 1954, creating a situation in which there were two Secretary-Editors during that month. This seemingly awkward arrangement was resolved by having the outgoing Secretary-Editor handle nearly-concluded business and the incoming Secretary-Editor handle on-going business. In the weeks prior to his departure, Mr. Wisely prepared schedules for several succeeding issues of the *Journal* to ease the burden on the new editor, who was moving into a position quite different from his previous experience. Because of the deadline pressure imposed by the monthly *Journal,* this assistance was important in maintaining the publication schedule. But even with this help, 1955 was a rigorous one for all members of the staff, as a survey of the activities of those first months shows.

Headquarters Move

At the same time that the Federation secretaryship changed hands, the Board of Control also decided to move the headquarters office to Washington, D.C. The Federation had grown to a point where it seemed that it could be more effective if the headquarters were located in a larger city, as many other technical organizations were finding. Chicago, New York, and Washington were all considered. W.J. Orchard, Chairman of the Federation's Finance Committee, favored Washington. He reasoned that the increased activity of the federal government in the water pollution control field, especially under the Water Pollution Control Act of 1948, made it important for Federation headquarters to be near the government. Orchard's recommendations prevailed, Washington was approved as the Federation's new headquarters by the Board of Control, and the Federation's Illinois office was closed after 14 years.

Although a national organization with strong international ties could publish a technical journal from virtually any location where good mail service existed, many members felt that the Federation had many more things to do in addition to publishing, and that a metropolitan headquarters was desirable. The move to Washington, D.C. was completed during the spring of 1955 with the new Executive Secretary-Editor and the Assistant Editor, Don Schweisswohl, the only staff members to move to Washington.

The Central States Water Pollution Control Association, one of the largest and most active in the United States, was disappointed with the move. Because of its pioneering achievements during the previous decades, that group was an excellent host for the Federation office. Joining with the Central States Association were several nearby groups that felt that the location of the Federation near the geographical center of the country was important. It was with hard feelings that the Central States Association gave up its physical proximity to the Federation Office, but the organization has since accommodated itself to the choice made by the Federation Board of Control.

The new location not only created access to the federal administrative and legislative processes affecting the field of water pollution control, but it also improved liaison with the printer of the *Journal,* Lancaster Press, of Lancaster, Pennsylvania. It became traditional to visit the printer at least once a year with new staff members so that there was close personal contact between the two offices, improving all aspects of *Journal* production and distribution.

When the Federation moved to Washington in 1955, 1,500 square feet of office space were leased in northwest Washington. This location served until March 1964 when a move to a new location five blocks from the original one increased that space to about 4,000 square feet with two internal expansions after the 1964 move. Further expansion continued in the years that followed and another move was made in 1976 to the present Federation location.

Participation of the Federation's elected officers increased greatly with the change in office location. During the 14 years in Champaign, only one President visited the headquarters during his term of office. After the move to Washington, each President visited the office not once, but repeatedly, and was far more involved in all aspects of Federation activities. Beginning in the late 1940's, elected officers were compensated for travel expenses, and their visits to headquarters definitely aided Federation operations and services.

Another advantage of the Washington, D.C. location was its proximity to New York for cooperation with such organizations as the American Water Works Association, The American Public Health Association, and the Water and Wastewater Equipment Manufacturers Association. The Federation cooperated in the publication of a manual of standard laboratory procedures, titled "Standard Methods for the Examination of Water and Wastewater." That publication received such acceptance that in many states it was designated as the official publication for legally recognized procedures of water and wastewater analysis.

From 1947, the Federation was a contributing member to this important volume published by the American Public Health Association and the American Water Works Association. With the tenth edition in 1955, the Federation became a full fledged member of the publication group and thereafter enjoyed all benefits as a publishing partner (See Chapter VI).

Federal Legislation

The first federal legislation authorizing $50 million a year in grants-in-aid for the construction of water pollution control plants was enacted in 1956. Feelings among the Federation membership ranged from the extreme right to the extreme left; some believed that a utility that must serve people indefinitely should be supported by the people without federal aid; others believed that such facilities were so expensive that they could only be built with federal aid. There were many persons who expressed the opinion that $50 million a year was an excessive amount for the federal government to contribute to the activity. Ultimately, federal contributions made up a great part of the support for such projects, and continued to grow. The commencement of this program gave the entire water pollution control field a strong impetus and led to accelerated growth.

Public Relations and a New Name

Because of this increased rate of growth in the field, many members felt that the Federation should begin a public relations program to focus on the importance of water pollution control. If such a program were successful, it would help all workers in the field with their programs and give them the recognition they deserved for their work.

Popular national magazines started publishing articles devoted to the quality of water throughout the country. Generally, these did not emphasize technical considerations, but they did cause the public to consider their local water supplies and water pollution control facilities. The discussions concerning starting a public relations program recalled the words of previous Secretary Wisely, who said, "the Federation will be wasting its money on a public relations program as long as the word 'sewage' is in its name." Throughout the 1950's, the Federation was known as The Federation of Sewage and Industrial Wastes Associations, a name which had resulted from much discussion in the 1940's (Chapter III). The Federation had three organizational names since its founding in 1928: Federation of Sewage Works Associations until 1950, Federation of Sewage and Industrial Wastes Associations from 1950 to 1960 and, thereafter, the Water Pollution Control Federation. The latter change was made during the terms of Presidents William D. Hatfield and Mark D. Hollis.

The last name change came about in this manner. In 1958, a name-change committee was appointed, headed by Roy Weston, with George E. Symons and Ralph Fuhrman serving as committee members. The committee reported to the Board of Control at its meeting in Dallas, Texas in 1959. Despite the previous consideration that had been given to eliminating the word "sewage" from the title, many members of the Board still felt no need for a name change, and no action was taken on the question at the first Board meeting at Dallas. At the second meeting, there was still a similar feeling, but when President Mark Hollis related his experience at a Dallas television station, a change in feeling occurred. The word "sewage" had been banned on the Dallas television station interviewing President Hollis, and it was therefore impossible for him to tell the name of his organization!

Although the time for a vote was near, many directors wanted to discuss the matter with their member associations before voting. Accordingly, it was agreed that a mail ballot would be taken within 60 days to determine the question. The resulting vote was in favor of changing the name of the Federation to its present name, the Water Pollution Control Federation. From that day forward, the Federation began to acquire a new image. The name change was effective January 1, 1960, at which time the name of the Federation journal became *Journal Water Pollution Control Federation*. By the end of the 1960's, articles appearing in U.S. papers and magazines were paying attention to water pollution and its control, although the lay press often confused the public water supply industry and the water pollution control industry. This confusion was not due to any activity of the Federation or the American Water Works Association, but it indicated the shallowness of perception of the writers (and many politicians) which was passed on to the public in many of their writings.

Another change of great importance was the recommendation by the Name Change Committee that the word "wastewater" be officially adopted by the Board of Control as a more generally acceptable word for "sewage and industrial wastes." The Board approved the term without hesitation and the Federation *Journal* started using it immediately; many other publications followed the Federation's lead. The word was suggested to the editor of

Webster's Second New International Dictionary for inclusion in its Third Edition, then under preparation. The suggestion was refused because the word was not in general use, but the word is found in Webster's New Collegiate Dictionary, published in 1973, acknowledging general acceptance of the term.

Many members were coming to feel that the name change was a major factor in the Federation's growth and that it demonstrated the desire and willingness of the organization and its leadership to keep pace with changing technology, to have due consideration for its public image, and to realize the importance of public awareness of the need for effective water pollution control. During these years the Federation developed a public relations and information program. Displays of Federation publications and brochures went to all U.S. Member Associations and a number of those abroad, and to numerous meetings of related organizations here and abroad. Annual conference publicity was distributed on both local and national levels by using all forms of news media.

Water Pollution Control Policy

At the annual conference at Philadelphia in October 1960, a most important step was taken when the Federation adopted a statement of policy on water pollution control in the U.S. During the course of the Board meeting, Morris Cohn, Sidney Berkowitz (Federation President 1967-68), Ralph Fuhrman, and others worked over early draft versions. Specific points of contention and argument included how to define pollution, and how to comment on "wastewater reclamation as related to the nation's total water resources." Some of the differences of opinion were not resolved, but an official statement was adopted. The complete statement as it was published in 1960 appears in the Appendix.

This statement has since been the basis of Federation positions taken before various Congressional committees conducting hearings on proposed federal legislation. It also has provided broad guidelines for the development of Federation policies and programs administered under the Executive Secretary. A prime example of its effect was the increase in the number of states requiring mandatory certification of wastewater treatment plant operators, a goal the Federation has sought for many years. In 1955, six states had such programs; in 1969, the number was 20, representing 57 percent of U.S. population.

The statement was widely distributed and clarified the Federation's positive attitude to water pollution control and problems. Over the years, as expansions and changes occurred in the water pollution control field, revisions were made in the statement to keep its impact current. The revision adopted October 10, 1974, is shown in the Appendix.

Honors and Awards

During this 14-year period of Federation history, four additional awards were authorized by the Board of Control. Full descriptions appear in the Appendix. The awards are:

- **William J. Orchard Medal** (1960) for distinguished service to the Water Pollution Control Federation.

- **Philip F. Morgan Medal** (1963) for originality, significance, comprehensiveness, effort, and verification of an idea by an in-plant study and solution of an operating problem.

- **Thomas R. Camp Medal** (1964) for demonstration by design or the development of a wastewater collection or treatment system, the unique application of basic research or fundamental principles.

- **The Gordon Maskew Fair Medal** (1967) for proficient accomplishment in the training and development of engineers, particularly at the graduate level.

Federation Growth

Throughout these years, there was persistent growth within Federation member associations as well as in the number of member associations. In 1955, a determined effort was launched to have all states represented by Federation Member Associations, and all states were represented by 1964. At the same time, overseas member associations were increased and strengthened, and the total number outside United States and Canada reached 18 by 1969. The further year-by-year growth of the Federation is evidenced by the data in Table IV-1.

TABLE IV-1. Federation Growth, 1955 - 1969

Year	Number of Member Associations	Number of Members	Annual Budget	Net Worth	Staff
1954	38	6 278	$100 300	$ 75 802	7
1955	38	6 782	103 000	81 394	
1956	38	7 102	112 250	90 191	
1957	41	7 759	121 800	93 551	
1958	43	8 323	123 200	86 245	
1959	44	8 675	152 700	77 244	
1960	44	8 513	168 000	108 032	
1961	45	8 843	182 550	139 825	
1962	47	9 364	194 825	155 559	
1963	47	9 917	204 200	138 786	
1964	49	10 830	263 200	93 369*	12
1965	51	12 083	343 000	89 472	13
1966	53	13 048	470 600	84 868	16
1967	54	14 272	535 000	119 282	17
1968	56	15 569	745 695	137 931	18
1969	57	16 573	792 650	145 370	23

The Federation was greatly interested in the formation of the International Association on Water Pollution Research (IAWPR), an activity which had its first conference in London in 1962, its second in Tokyo in 1964, and its third in Munich in 1966. During these years, the Federation offered to represent the U.S. in IAWPR because of its experience, interest, and strong national representation in the field of water pollution as well as water pollution research. The offer was not accepted. Instead, the Federation became one organization of several to represent the U.S. through a national committee. Through this activity, the Federation became better known by its exhibits at IAWPR conferences and for its water pollution control interests around the world. The United States and the Federation were looked to increasingly for technical knowledge, and the growing international reputation of the Federation led to the development of overseas member associations.

---

* Decrease in net worth caused by rising costs, particularly paper and printing; and by funding of new programs during presidential terms of Messrs. Seidel and Steffen.

During the period ending in mid-1969, the Federation staff grew from 7 to 23. This growth reflected an increase in staff work to support additional committee activities, to maintain larger mailing lists, and to keep in touch with the general increase in water pollution control activity in the U.S. A prime example of a major addition was the initiation of the publication of the Federation newsletter, "Highlights," authorized by the Board of Control at its 1963 meeting in Seattle. The first issue was one containing the speech given by John Charles Daly at the Seattle Conference. Regular publication was started in January 1964 and was continued on a monthly basis. The newsletter provided a means of contacting the membership more quickly than was possible through the monthly *Journal,* with its necessarily long press time.

Technical Services and Committees

With the staff reorganization in 1965, technical services were greatly expanded. This program focused on developing safety programs for wastewater collection and treatment facilities, promotion of operator training and certification, and servicing inquiries from a wide range of interests, from school children to technical personnel.

The Federation's committee structure grew along with the organization. Hundreds of members voluntarily worked in the many and varied committees with appropriate and detailed staff liaison. Areas of activity included annual conference programs, industrial wastes, research, personnel advancement, technical practice, wastewater reuse, safety, standard methods, publications, legislation, manpower needs, public relations, and various joint activities with other organizations having related interests.

A significant change in the committee structure of the Federation came in 1956, when the principal standing committees with constitutional status were reorganized with the Chairmen appointed for a limited term of 5 years. The tabulation of committee chairmen given in the Appendix shows the changes at that time. The Appendix also contains a history of selected constitutional committees.

Anyone acquainted with Federation activities knows that the organization is heavily dependent on its committee members for its work. The time the members spend on Federation work is probably equal to that spent by the staff personnel. The numerous manuals of practice which were actively revised and extended during the years 1945 to 1969 are indicative of this work, as are the various editions of "Standard Methods," and the annual research committee effort on the review of the literature.

Annual Conferences

Another measure of organizational growth was the increase in size of annual conferences. Statistics published annually in the *Journal* give complete data, but here it might be noted that the Federation first crossed the 1,000 mark for conference registration in 1952, the 2,000 mark in 1961, the 3,000 mark in 1965, and the 4,000 mark in 1967. The registration for the last conference of this period of Federation history, the 1968 Chicago Conference, was 4,806. The extent of the technical program also grew through the same period as did the number of papers, number of sessions, and diversity of those sessions. The 1955 conference had nine sessions with 21 papers, and by 1968 the conference had grown to 25 sessions and 114 papers. Finally, there was a major increase in the commercial exhibits at the conference, that were managed throughout this period by the Water and Wastewater Equipment Manufacturers Association.

Administrative Changes

A major organizational change occurred in 1967 when the Board of Control created the office of President-Elect. This provision distributed more evenly the various duties of the officers, particularly with respect to visitations to Member Association meetings. During Fuhrman's tenure he visited, as Federation representative, each of the existing 56 Member Associations in the U.S. and abroad.

One Federation group deserving recognition is that of the Presidents who filled the office, served for a year, and then returned to other activities. Each of these men had his special interest, concern, and suggestions as to what the Federation should do in one direction or another. To act on these suggestions over a period of years could have caused tension between presidential expectations and staff capability, and the on-going review by the

Executive-Secretary of Federation programs and their effect. This constant evaluation of staff activities and elected officers created a minimum of conflict or confrontation, and on the whole, there was mutual acceptance and a maximum level of production between officers and staff. The Federation Presidents who served during the period were highly talented men, all dedicated to a major organizational effort for better water pollution control. Full recognition of individual contributions is not possible, but the accomplishments of the Federation are reflections of their efforts.

Dr. Fuhrman left the office of Federation Executive-Secretary on June 30, 1969, to become Assistant Director of the National Water Commission. He had served the Federation for nearly 15 years. At the time of his departure, he said that he left with regret but with a feeling that continued progress in the organization was assured.

Robert A. Canham was appointed by the Executive Committee to the position of Executive-Secretary pro tem as of July 1, 1969. He had joined the Federation in 1957 as Associate Editor of the *Journal*. In 1964, he was named Editor of the *Journal* and Assistant Executive Secretary in recognition of his responsibilities, thus freeing the Executive-Secretary for duties as the Federation's Executive Officer. Canham was the obvious choice to become the fourth Federation secretary. The change was a smooth transfer of leadership in the headquarters office, and marked the end of the third period of the Federation's first half century.

On February 29, 1964 the Federation moved its headquarters to 3900 Wisconsin Avenue, N.W., Washington, D.C. 20016. Located on the top floor in one wing of the Williamsburg-style building, the Federation's office space was nearly doubled. Expansion was made several times during the Federation's 12-year residency.

As the 10,000th member of the Federation, William C. Henry, Assistant Chief Engineer for the Wilmington, Del., Department of Public Works, is shown above (3rd from left) receiving a membership certificate from William Hokanson, President of the Maryland-Delaware Water and Pollution Control Association. Dr. Harris F. Seidel, WPCF President, and Dr. Ralph E. Fuhrman, WPCF Executive Secretary, witness the ceremony.

Presidents and Executive Secretary during 1955-69

D. B. Lee
1954-55

F. W. Martin
1955-56

E. C. Jensen
1956-57

K. S. Watson
1957-58

W. D. Hatfield
1958-59

M. D. Hollis
1959-60

R. E. Lawrence
1960-61

H. E. Schlenz
1961-62

J. E. McKee
1962-63

H. F. Seidel
1963-64

A. J. Steffen
1964-65

R. S. Shaw
1965-66

A. D. Caster
1966-67

S. A. Berkowitz
1967-68

P. D. Haney
1968-69

R. E. Fuhrman
1955-69

Philip F. Morgan for whom the Philip F. Morgan Award was established.

Thomas R. Camp for whom the Thomas R. Camp Medal was established.

Gordon Maskew Fair accepting from President Paul Haney the first Gordon Maskew Fair Medal for his distinguished service to the field.

Chapter Five

Rounding Out A Half Century-- 1969-1977

In the words of then President Paul D. Haney, the year 1969 was an "End and Beginning." At the end of June, Dr. Ralph E. Fuhrman completed almost 15 years of dedicated and effective service to the Federation as Executive Secretary; Assistant Secretary-Editor Robert A. Canham was chosen at that time to fill the position. During the interim from Fuhrman's resignation to Canham's appointment by the Board of Control at its Dallas meeting in October, the latter served as Executive Secretary pro tempore, having been appointed by President Haney with the approval of the Executive Committee.

In the period just ending there had been many advances in Federation programs and capabilities, including: a greater variety of publications (in addition to the three periodicals); expanded technical services supported by diversified professionals; an expanded and strengthened international posture; and repute in the international technical community. The era just beginning brought a challenge to the new Executive Secretary, to the officers, and indeed to the entire membership, to meet rapidly increasing needs in the water pollution control field.

This new period could not have begun in a more appropriate place than Dallas, Texas. It was there, only 10 years earlier, that the Federation had changed its name to the present one, and took other steps to improve its image and meet the challenge of providing the knowledge and leadership needed in the water pollution control field. Perhaps the best vantage points from which to view the impacts of these changes are those of the presidents that have served the Federation since then. Thus, the next several sections of this chapter will describe the Federation's activities and growth as discussed in each of the presidential messages published annually in the September issues of the *Journal* during this period.

Paul D. Haney (1968-69)

At the Dallas Conference in 1969, WPCF President Haney pointed out the unique structure of the Federation—56 autonomous regional, state, or national water pollution control associations in the U.S., Puerto Rico, and 16 other countries. He contrasted the Federation with other technical and professional organizations, such as the American Society of Civil Engineers and the American Water Works Association, both centrally organized. Haney emphasized the word "autonomous" as the key to the Federation's strength, because of the initiative it encourages each Member Association to assume. Another strength is provided by the broad range of interests represented by Member Associations—teachers, researchers, managers, operators, regulatory agency personnel, consulting engineers, industrial wastes personnel, manufacturers representatives, students, and the interested citizen. As each of these interest groups directed the Federation to their particular concern, many types of problems had to be encountered, and the Federation became a viable organization. The Federation's *Journal* and special publications along with the numerous publications of the Member Associations aided in furthering the efforts of the various bodies and of the Federation itself.

President Haney suggested the great need for the Federation to become more involved in promoting better management of wastewater utilities and resolving many problems associated with plant operation. "Too frequently," he said, "communities commit themselves to long bond issues in order to construct a wastewater facility, then through lack of qualified management and operation, do not receive benefits commensurate with expenditures." He pointed out that the benefits of the Federation operator training courses, training grant program, manuals of practice, and safety and operator certification programs were important contributions to improving plant operations but said that comparable work in management was sorely needed.

Joseph B. Hanlon (1969-70)

President Hanlon, in reporting to the membership in 1970, projected current trends into the future. He reminded the Federation of the axiom stated by the noted newsman, and honorary Federation member, John Daley, in his October 1969 address at the Dallas Conference: "An effective program of pollution control must not trickle down from the top but must bubble up from the interest and commitments of the public." Hanlon pointed out that during 1970, citizen interest in ecology and the environment became widespread. The Federation finally had great support for its fight against water pollution.

In 1970, the Federation began to become aware of the increasing responsibility placed on its membership by this developing public interest. That year, Executive Secretary Canham's report to the Board pointed out the many increasing activities by laymen and the increasing requests to the Federation for information about water pollution control. He emphasized the additional responsibilities placed on the technical community, especially the Federation. He also pointed out that despite the increasing interest on the part of the public during the past several years, during 1970 more sewer bond issues failed than had failed for many years. Part of the cause was the impact of higher interest rates, but in many instances it was public dissatisfaction with technical and administrative leaders.

Executive Secretary Canham's 1970 report also pointed out that pollution became an issue on American campuses that year and was debated loudly, and often in the same context as other issues, such as the Vietnam War. Student frenzy reached a peak in April with an Earth Day that turned out to last several weeks. Federation members participated in a number of Earth Day activities and found student concern to be both widespread and encouraging. Many confrontations developed. There was a persistent minority of vocal social scientists, more often faculty members than students. It became apparent that there was practically no communication between engineering and social science faculties. In many universities, the engineering schools showed no desire to develop a dialogue with the social sciences. Canham emphasized the need for the Federation to develop a change in its attitudes and direction in order to satisfy new social demands.

Along with the social upheaval during that year, changes in the federal government culminated in an executive order creating the Environmental Protection Agency. This significant move placed nearly all environmental responsibility in one federal agency under the jurisdiction of the White House. The Federation endorsed the concept of the EPA but was concerned that creation of the new agency did nothing to consolidate several agencies. One of the benefits to be derived from putting all environmental programs in one agency was a better coordination of efforts in water supply and pollution control. Pollution control technology, Canham said, "can produce water of drinking water quality. Nearly 150 plants with this capability now are in various stages of design and construction. . . . Treated wastewater was introduced directly into drinking water distribution systems in locations outside the U.S. and it is predicted that the time is near for a test site here." (See Appendix for "Joint Resolution of the AWWA and WPCF on Potable Reuse of Water.")

One incident that occurred in 1970 serves to demonstrate the increasing ability of the Federation to mobilize itself for effective action. In that year, the Federal Water Pollution Control Administration, at the time the agency in the U.S. Department of Interior that had overall control of the federal water quality effort, announced plans to hold a federally-sponsored conference similar in nature to, and scheduled to be held shortly before, the annual WPCF conference. Such a move could have seriously affected the success of the Federation conference, and the Federation and its Member Associations, under the strong leadership of President Joseph Hanlon, strongly expressed their opposition, through appropriate government channels, to the FWPCA meeting. The effort was successful, and the meeting was cancelled. One interesting sidelight to the incident, perhaps more than a coincidence, is the fact that two weeks before the Federation conference the Assistant Secretary of the Interior responsible for water pollution control programs resigned.

Art Vondrick (1970-71)

In his address to the membership in 1971, Art Vondrick said: "Almost every Federation committee has been affected by the high level of federal activity in water pollution control. The amount of correspondence and other contact between committees is tremendous and still growing. Government Affairs could not be a more appropriate name for that committee, but it would seem equally appropriate to give assistance to many other committees because all have become involved in government affairs in one way or another. A considerable part of all Federation affairs is involved with, or a direct result of, the federal government and its actions . . .

"Contrary to the former style of committee endeavor, no one seems to have control over his own deadlines any more. Whereas in previous years, almost any committee could rely on its yearly instructions from the Board of Control as its chore, it is not unusual for many committees to have their work increased by virtue of rapidly moving events or be forced to review their work to see if they are in step with the times. The day of the committee chairman's doing a lion's share of committee work is far behind us. The input from active committee workers is most valuable and important." Vondrick was "gratified by the response to a letter that was circulated to all Federation Directors emphasizing the need for greater involvement on their own part as well as that of members of their respective Member Associations." Even a year later, names of volunteers for working committee assignments were still coming in.

Vondrick added that "as the Federation officers visit Member Association meetings, there seems to be an emergence of the Federation's stature in the making. We have done a lot of talking to ourselves in the past two decades because no one else would listen. Now hardly a month passes when a request is not received to participate in a meeting or in an event that few of us knew existed, or at least cannot remember . . . It is a definite challenge. We have more opportunities than ever to tell our story. As a group we will be relied on to practice what everyone is preaching, and we will have to do a little preaching too. This is not much different from the way it has always been, except that probably more people with influence will begin to believe us. To the ordinary citizen and city official, water pollution controllers did not necessarily create an aura of excitement. We meant well and we were sincere but, after all we either had a vested interest or we were impressed with ourselves. All of a sudden maybe we were right after all, but it took passage of some legislation by the Congress to attract people to the cause. The years ahead for the water pollution control industry and the Federation are difficult to predict."

Joseph Lagnese (1971-72)

In his 1972 message, Joe Lagnese departed from the practice of looking back on activities and pointed to particular needs of the Federation in the future. "Of major concern to the organization now", he said, "must be the consideration of whether the public and the Congress will continue to maintain their interests in water pollution control to the extent of the recent past or present. It is no great revelation to acknowledge that the Federation in the past few years has experienced unprecedented growth and has attracted national interest and attention to a greater extent than we could have imagined previously. There should be no doubt that this newly attained aura of prestige and popularity is attributable in large part to the concurrent increased public and governmental interest in our work. This is only reasonable; and it should also be reasonable to assume that the Federation in years ahead is going to continue to be influenced and affected by these same factors.

"Although it is difficult to predict the future attitudes and values of the general public, we can only be guided by our own convictions that water pollution control is essential and that it must remain a high priority commitment of both people and government." Lagnese added: "The Federation has no other choice and in effect has an obligation to plan for a future in which water pollution control will remain a critical need for all nations of the world and that the organization will remain a useful contributor to the satisfaction of this need."

At the time it seemed unlikely that all future issues could be anticipated, but he believed that the Federation could reasonably prepare itself by continually and realistically appraising capabilities and developing alternatives to those found to be less than desired. Lagnese called for continued Federation involvement in government water pollution control activities and disagreed with the contention that following enactment of the 1972 amendments to the water pollution control act, these activities should be minimum. He pointed out that the organization's role should not be limited to reaction but that, "we should now strive to be effective in assuring that Congress remains interested in water pollution control programs of the nation" and "that Congress be kept fully cognizant of the defects of their contemplated actions and the need for legislative adjustments and innovations."

Floyd Byrd (1972-73)

In September 1973, President Byrd said, "Environmental improvement is an idea whose time has come and the Federation must play a role in developing this ideal on rational, practical, and effective lines. Any scheme for improving man's environment can be badly warped and distorted without the help of such professionals as the members of the Federation." Byrd said that during his administration, nothing radical had been initiated, but the trends and directions initiated by his predecessors had been continued. He also pointed out that "fortunately the Federation is a democratic organization and changes can be made only as fast as its members can agree to such changes." From his visits to more than 25 Member Associations in three years, he reached the conclusion that "the thrusts and directions the Federation is taking meet the general approval of the membership." To him "the beauty and the strength of the Federation lies in its many relatively independent and autonomous local associations," which "have nearly complete freedom and power to run their own affairs at the local level in order to meet local needs, with little interference from Federation headquarters."

John Parkhurst (1973-74)

In September 1974, President Parkhurst reviewed some of the Federation's past efforts that had been nonproductive and suggested that there was a "growing consensus that even qualified support of Public Law 92-500" might become such an effort. The law, which was passed over Presidential veto, "has fallen far short of its promise in the eyes of even its most ardent supporters," and "at issue is enhancement of water quality commensurate with the amount of public funds committed." In citing the questionable wisdom of imposing a uniform requirement of secondary treatment, Parkhurst said, "extensive marine biological research has not revealed a deleterious effect on ocean organisms caused by oxygen-demanding and nutritional wastewater constitutents, provided that properly designed outfall diffuser systems are used in the open ocean. However, unless corrected by legislative amendment or administrative interpretation, PL 92-500 will commit billions of dollars of public funds to secondary treatment of waste before marine disposal, with little or no significant improvement in marine waters expected."

Parkhurst added, "Equally questionable are policies that arbitrarily ban the disposal of organic and other materials into the ocean and provide for alternative practices that may create serious environmental problems or bring about the eventual waste of energy resources. This is not to imply that ocean disposal should not be critically examined, as it is now being done under the auspices of the Southern California Coastal Water Research Project, nor that the ocean is insensitive to waste discharges. What should be examined, however, is the legislative fiat that has made mandatory a specific method of treatment without consideration for the relevant needs of the receiving waters."

Parkhurst reported on the seminars and workshops conducted by the Federation in cooperation with its Member Associations and the EPA. These workshops were held in each region of the U.S. during 1974 to explain the provisions of the recently enacted PL 92-500 and how the EPA visualized their implementation. These workshops were highly successful in bringing about extensive discussions among people responsible for developing and enforcing the regulations under the act and people responsible for complying with these regulations. They produced a long list of concerns about certain impractical provisions in the law. Highlights of the discussions of the various regional workshops were published by the Federation and widely distributed, with the hope that the concerns and suggestions recorded would be heeded by regulators and legislators responsible for making interpretations and any necessary adjustments in the act.

Sam Warrington (1974-75)

In September 1975, President Sam Warrington recognized that the Federation from the beginning had been a technical organization that had conducted a steady but highly conservative program serving its members through its *Journal,* other publications, and annual conferences. Warrington pointed out that prior to 1956 there was no substantial federal legislation to interpret, and as a result there was little controversy among the local authorities, industry, the states, and the federal government. With enactment in 1956 of the first permanent water pollution control legislation, which included federal subsidies for construction of public wastewater treatment facilities, the rate of water pollution control activity accelerated and relationships among the various groups in the field began to change. The resulting accelerated activity also brought a rapid increase in associated problems.

Warrington pointed out that "Federal legislation has been expanded and revised several times since 1956, culminating in Public Law 92-500." The federal program has grown from a very small number of people in 1956 . . . to an agency with nearly 10,000 employees. Very few of the 10,000 Environmental Protection Agency personnel are professional in a technical sense, and this evolution should tell us something about the different approaches now needed. As a result, the Federation had to make adjustments to unfamiliar circumstances. Warrington added, "While the Federation remains a technical organization, and there are an increasing number of reasons to emphasize this, it has changed from a passive to an activist organization. This change has not been easy but it is beginning to show signs of satisfying both broad and specific needs."

Victor Wagner (1975-76)

In September 1976, President Victor Wagner reviewed addresses of previous Presidents commencing with Harry Schlenz in 1962, and concluded that "the results of involvement have nowhere been more apparent than in the area of public affairs." Wagner pointed out that the gradual increase in public affairs involvement prepared the Federation for its important role as related to PL 92-500. Based on information produced by two series of workshops on the law, its requirements, implementation, and adjustments needed, the Federation prepared its recommendation for improving the law (Chapter VI). These recommendations were used as the basis for presentations to the Congress and to the National Commission on Water Quality.

The National Commission on Water Quality, created by PL 92-500 to study all aspects of achieving or not achieving the 1983 goals of the act, submitted its report to Congress March 18, 1976. Following that report, the Federation and its Member Associations responded with the 10th annual government affairs seminar held on April 6, 1976, with the theme "Public Law 92-500—Mid-Course Corrections." This seminar attracted a record attendance by members of Congress and their staffs and included presentations by representatives of the U.S. EPA, staff members of the Senate Public Works Committee, and the House Committee on Public Works and Transportation and the National Commission on Water Quality.

Horace L. Smith (1976-77)

On assuming the Federation presidency on October 7, 1976, Horace Smith told the Board that a review of Past President Wagner's message in the September *Journal* "disclosed an outline for the continuing development and progress of programs and projects of the Federation." Smith went on to speculate on "some current and specific needs," which appear in his Epilogue to this history (Chapter VIII). President Smith offered these needs as a WPCF program for water pollution control and went on to say, "The escalation of pollution abatement legislation and regulations and the resulting demand for the accelerated advancement of the state of the art of water pollution control has made it most difficult to keep the components in perspective and balance. This will be our challenge, and responsibility, for the next several years until the degree of change has stabilized."

Admitting that, "The equation cannot be balanced by emphasizing particular factors to the exclusion of others" President Smith continued, "we must recognize priorities as a matter of practicality." He listed program areas to receive attention in 1976-77 as: "membership development and service; Member Association affairs, both national and international; aims and objectives; Federation organization; assessment of current states of the art and technology; technical practice; manpower planning and training; research emphasis; cooperative efforts; and organizational management strategies for WPCF programs."

Describing plans for several of these programs in October 1976, President Smith said he did not believe "that the Federation membership is happy with the defensive role of responding, or reacting as the case may be, to legislative and regulatory actions of the Congress and of the Environmental Protection Agency. Considering the expertise contained within the membership of the Water Pollution Control Federation through its Member Associations, we should be able to develop a comprehensive legislative and regulatory strategy of wastewater and water pollution control which: is highly responsive to the protection of the environment and balanced with technology; is sensitive to urban, industrial, agricultural, and silvicultural activities; is flexible and cost-effective; is progressive in its implementation; and which minimizes administrative constraints and delays. Such a strategy should recognize local options and prerogatives and should not have fragmentation of responsibilities. I have asked the Government Affairs Committee to pursue the development of such a strategy for the Federation to utilize as an aggressive approach for our inputs to EPA and to Congress."

Smith also pointed out that changes in technology had exceeded the Federation's ability to produce the manuals needed to present this technology, and that, "the Technical Practice Committee has concluded that the solution to this problem is the production of a greater number of manuals of practice relating to more specific or specialized subject matter than the traditional complex and comprehensive manuals of the past. The selection of subject matter for these mini-manuals needs to reflect upon several factors; e.g., purpose, users, management systems, system functions, organization activities, system analysis, and even further into the various and numerous subfactors." For the Federation's 50th year, President Smith viewed "the work of the Technical Practice Committee . . . as a most important activity of the total effort of the WPCF, if not *the* most important activity in terms of our credibility in projecting ourselves as the foremost organization representing the expertise in the field of wastewater and water pollution control."

Thus, each president since 1969 analyzed the Federation and its objectives and progress toward those objectives and contributed uniquely towards improving the Federation. One measure of the changes that occurred in this period was the constantly increased amount of time and involvement by Federation Presidents. Over the years, most Federation members were unaware of the increasing daily responsibilities that Federation Presidents took on, even while maintaining their full-time technical careers. Great appreciation is due each President for his contributions and increased involvement in Federation affairs.

Changes and Activities, 1969-1976

One of the first changes in 1969 and 1970 was a reorganization of the staff. For many years, primarily because of its small size, multiple duties had been assigned to some of the staff. As the Federation and the demands for services and activities grew, it became evident that a division of responsibilities and a new identification of activities were necessary.

For many years the Executive Secretary was also the *Journal* Editor. That arrangement changed in the early 1960's when the second position on the staff was changed to Assistant Secretary and Editor, thus allowing the Executive Secretary to spend full time in administrative duties.

Shortly after the change of Executive Secretaries in 1969, the staff was reorganized to provide an Assistant Executive Secretary and a separate position of Editor. The position of Assistant Executive Secretary was filled by Leo Weaver who, for three years, spent a major part of his time building an effective public affairs program at the national level as well as carrying on additional administrative duties. When Weaver left to become Executive Director and Chief Engineer of ORSANCO, the Assistant Executive Secretary position was filled by Robert Perry, who came to the Federation with an extensive background in engineering and administration.

When the changes were made in 1969, Bob Rogers became Editor; he had been Associate Editor for several years. When he left the Federation about two years later, Peter Piecuch became Editor.

In the eight years following 1969, the editorial staff doubled and in 1977 it numbered 12 persons, the largest group within the staff. The earlier practice of employing engineers or scientific professionals on the editorial staff continued, and in 1977 it included four engineers and/or scientists.

The general staff reorganization included the division of activities into seven major programs: Editorial; Technical Services, headed by George Burke; Public Affairs, headed by Ken Kirk, the first lawyer employed by the Federation; Education, headed by James Suddreth; Membership Services, headed by Jack Wilson; Public Relations, headed by Philip Ridgely; and Office and Conference Management, headed by Robert Dark. Changes in business practices and responsibilities also led the Federation to increase its use of outside legal counsel for a variety of matters.

Growth of the Federation and its services during the period 1969-1977 is shown in Table V-1.

Leonard Rossi, a wastewater treatment plant operator, from Arvada, Colo., was honored as the Federation's 20,000th member in 1973. WPCF President J. Floyd Byrd (right) presents a special membership certificate as Rocky Mountain WPCA President Farrell McLean (left) looks on.

Paul J. Thompson was honored as the 25,000th WPCF member during its 1977 Government Affairs Seminar. Federation President Horace Smith (right) presents the certificate to Thompson, who is in charge of a regional planning and development commission, a profession that exemplifies WPCF's broadening interest in other fields.

TABLE V-1. Growth of WPCF, 1970-76

Year	Member Associations	Members	Annual Budget	Net Worth	Staff
1970	55	17 628	$ 942 573	147 830	28
1971	58	19 254	1 012 423	137 455	
1972	59	20 212	1 207 400	341 176	33
1973	63	21 540	1 398 995	509 839	37
1974	63	23 118	1 687 210	697 986	38
1975	63	24 497	1 879 800	755 920	40
1976	65	25 335	2 573 100	737 758	41

In May 1976 the Federation moved to its present headquarters at 2626 Pennsylvania Avenue, N.W., Washington, D.C. 20037. Located on the outskirts of the Georgetown section of the city, the building is 10 blocks from the White House.

Headquarters Location

Offices for the staff changed several times in the post-1969 period. When the Federation moved to Washington, D.C. in 1954, it was located at 4435 Wisconsin Avenue. Ten years later, it leased about 2,985 square feet on the fourth floor of a handsome Williamsburg style building at 3900 Wisconsin Avenue. This building, the home office of the Equitable Life Insurance Co., was situated on nearly 10 acres of beautifully landscaped property in the northwest section of the city.

As the staff grew, more space was added several times until 1975, when the Federation occupied about 1½ floors of the north wing. Late in 1975, and with little warning, an eviction notice was received. The insurance company had sold the property and, with a 6-months' cancellation clause in the lease, there was no choice but to move. So, early in May 1976, the Federation headquarters moved to a new building at 2626 Pennsylvania Avenue, N.W., owned and partially occupied by the National Telephone Cooperative Association. The site is near the Potomac River, near Georgetown, 10 blocks from the White House, and is surrounded on three sides by National Park Service property.

Budgeting and Programming

Facing demands for increasing programs and services beyond the increase in its financial resources, the Federation became aware of the need for more scrupulous budgeting and accounting for these limited resources. During his administration, President Joe Lagnese pointed out that one of the most potentially fruitful efforts that could be pursued to ensure effective utilization of resources would be the revision of budgeting and accounting procedures. He urged a move toward cost center budgeting.

In simplest terms, the budgeting and accounting procedures long used by the Federation had categorized costs according to general expenses, such as travel, supplies, or salaries, rather than according to how these expenses were being used to accomplish specific programs. With this procedure, it was difficult to establish priorities for financial commitment to any one of the many programs that the Federation might consider. This situation existed because the procedure does not provide for the apportionment of various purchase requirements into specific program needs and because it does not provide budget controls that would ensure that each program receives the intended amount of support in terms of all purchases. In effect, that system only provided for new programs and improved efforts by increasing the allotment for general purchases, such as total staff wages or total travel. The system had a built-in trust that all programs for the ensuing year would receive expenditures and staff effort in proportion to that intended by the original program planning. Although the system had been genuinely workable in previous years, the factors involved, as the Federation grew, were too large and too complex to rely on such a minimum control system to allocate limited financial resources. Therefore, program-oriented budget development and control became necessary. The more rigorous cost accounting suggested for the Federation was not a simple matter to develop and administer, but continuation of the purchases-oriented budget and accounting procedures was no longer justified.

The changes were a logical outgrowth of the financial growth of the Federation. In 1969-70, the budget was $ 942,573 and the reserves of $ 147,830 were quite low in comparison with the operating budget. As it became obvious that the growth rate would be increasing rapidly, intensive planning was undertaken. In 1971-72, along with the budget growth, the reserves began to improve and by 1976-77 stood at about $ 800,000. This, however, was less than the 50 percent of budget level approved as the objective by the Board of Control.

Along with growing reserves was the need to make the funds work, and as a result a continuing investment program was developed and maintained. For most of the recent years of the period, the reserves were in short-term government, quasi-government, or bank certificates of deposit and the yield was a significant part of the operating budget.

Budget control also continued to be a problem, and as its complexity increased so did the monitoring and supervising required. For several years an outside contractor was used to provide accounting records on an automated basis. Following the acquisition of computer facilities in 1975, plans were started to computerize the entire financial operation by early 1977.

Public Affairs

The impact of the federal legislation in 1956 and subsequent amendments had a profound effect on the Federation. The beginning of the construction subsidy program in 1956 essentially coincided with the beginning of a new era of steady, though not dramatic, growth and involvement in legislative affairs by the Federation.

A large number of Federation members became involved when the legislation was being considered in 1971 and 1972 and even larger number were involved following 1972 in a series of workshops throughout the country that led to a statement on the implementation of the Act. The points in the position statement were used and referred to repeatedly by the Congress, EPA, and the National Commission on Water Quality as needed corrections in the law became evident.

During the last eight years of the Federation's first half century, there was continuing and intense participation in Congressional hearings and regulatory processes. Position statements and testimony were developed through an extensive consensus procedure and, as a result, statements made always represented the Federation rather than any individual member's view.

Many problems in the implementation of PL 92-500 became evident and the Federation carried on a continuing program to contribute technical expertise to solutions to these problems. Challenges seemed to multiply—problems in the huge construction program in the public sector, planning problems, permits and monitoring requirements, enforcement, and public involvement. During this time the Federation was one of the leaders in the non-government sector.

Conference Exhibits Management

From the beginning, equipment exhibits at the Federation's annual conference were managed by the Water and Sewage Equipment Manufacturers Association (now the Water and Wastewater Equipment Manufacturers Association, WWEMA). For most of these years, little or no contractual arrangement existed between the two organizations, although from the beginning WWEMA contributed part of the income from exhibit space sales to the Federation (Chapter VI).

WWEMA contributions prior to 1969 were: 1940, $1,500; 1941-48, $5,000 per year; 1949-59, $6,000 per year; 1960-64, $7,500 per year; 1965-68, $10,000 per year. As WPCF activity increased, the need for having a more definite understanding became necessary. In the absence of an understanding on what portion of the exhibits income would be transferred to the Federation, no definite planning could be made. In the early 1970's, intensive discussions among representatives of WWEMA and the Federation focused on developing a contract. These resulted in an arrangement for the 1972 and 1973 conferences.

During the later years the Federation began to realize the importance of managing the exhibit program for its own Conferences. After a series of discussions, the Board of Control in 1972 voted to end the exhibit relationship with WWEMA. Beginning in 1973, staff was added to carry on the function, and the first Federation-managed exhibits were in Denver in 1974. Thereafter, the exhibit program was managed by one Federation Exhibits Manager and an assistant. The growth in exhibits was immediate and continued rapidly in 1974-76 to a point where only 10 or 12 cities in the county could accommodate the Federation Conference. Exhibit activity and income are shown in Table V-2.

TABLE V-2. Growth of Federation Exhibits

Year	Number of Exhibitors	Exhibit Area (sq ft)	WWEMA Contribution to WPCF
1969	116	22 400	$25 000
1970		25 000	25 000
1971	136	30 000	61 750
1972	150	37 000	80 640
1973	168	44 000	93 385
1974	200	49 500	
1975	223	55 900	
1976	276	61 800	

The financial impact on the Federation budget during these three years was sufficiently strong that no dues increases were necessary.

Membership Services

Prior to 1971, membership records were maintained on a device that, although semi-automatic, required considerable manual handling. Cards were punched with member's name, address, and additional information regarding member association, class of membership, and date of dues payment. By the use of a device containing a fluid, names and addresses could be transferred to envelopes or wrappers for the mailing of publications. A contract was signed with a local computer service bureau on May 11, 1971 to convert the membership records to a computer maintenance system and 19,058 records were converted in June 1971. Monthly record updates were performed during the computer service period.

As potential uses of the computer became more obvious, more flexibility was built into the system, allowing the Federation to offer member listings and mailing labels to Member Associations at direct cost. Originally, approximately eight listings were sent by the Federation to the Member Associations each year. This was later expanded to once monthly.

Starting with the 1972 membership year, WPCF offered a centralized billing service to Member Associations in the United States and Puerto Rico. The intent in providing this service was to speed up the renewal process for existing members by having the dues payments sent directly to the Federation. Because of the capability of the computer system, the Federation was able to offer this service to associations at a relatively low cost. The new system saved much time for individual association secretaries because it was not necessary to prepare renewal notices and receipts for payments did not have to be recorded and forwarded to the Federation. Fourteen associations participated in this service in 1972, when there was a 50 cent charge for each renewal received directly by the Federation. In October 1972, the Board of Control directed the staff to extend the centralized billing service at no charge to all associations in the U.S. and Puerto Rico. For the 1973 membership year, 31 associations elected to participate in this service. The service was expanded later in 1973 to include the issuance of a membership card to members of associations that participated in the direct billing service. Membership cards were offered to the associations as a further free option if they were participating in the direct billing service.

After 1972, the alphabetical member listing in the *Journal's* Directory issue, published in March of each even-numbered year, was produced by using the computer records as source material for a photo-composition machine.

In 1974 the Federation offered Member Associations the opportunity to establish a special classification of association member for persons who desired association membership but not the full privileges of Federation Active Members. Though no classification name was officially adopted, this class became known as "affiliate." On payment of a $3.00 fee to the Federation in addition to association dues, affiliates receive monthly from the Federation a copy of *Highlights/Deeds & Data* and are included on the membership listing which is furnished to the associations monthly. Affiliates are eligible to purchase Federation publications at member prices. When the association chooses Federation direct billing, affiliates are billed along with other members and rebate of association dues is made to the association on a monthly basis. By the end of 1974, approximately 100 persons were recorded on Federation records as "affiliates"; by the end of 1976 there were over 1,200 affiliates.

Maintenance of membership records was performed by the computer service bureau until October 1975, at which time the Federation arranged for rental of an in-house computer. Some severe problems were experienced in the early months of in-house membership record maintenance. In time, the problems were smoothed out and more responsive and up-to-date membership information resulted.

By 1976, membership services offered to member and affiliated associations in the U.S., Puerto Rico, and Canada included:

- Direct mailing of renewal notices to members with self-addressed return envelope;
- Receipt, processing, and tabulation of renewal payments and monthly rebate of association dues;
- Monthly listing of each association's membership;
- Monthly issuance of membership cards mailed directly to the members;
- Inclusion of dual membership and affiliates on the membership listing;
- A monthly reconciliation listing indicating all activity in that association's membership in the preceding month;
- Capability of producing membership listings for an association categorized by "section" within that association, if that information has been submitted to WPCF;
- Printing, stocking, and engrossing of membership certificates at the association's discretion.

In 1976 the following services to associations were added:

- Computer service products at cost, such as pressure-sensitive labels, cheshire labels, and rosters in any sequence;
- Provision for the addition and maintenance of non-WPCF membership on the computer system. This service includes records of any individual not ordinarily on the WPCF membership rolls, but who will, at the direction of the association, be maintained for the purpose of producing labels and/or listings for an association's use. This service, offered at WPCF cost, was introduced in 1977.

Member Association Regional Meetings

At the Atlanta meeting in 1973, the Board of Control approved a budget that included a new program called Member Association Regional Meetings. This program called for participation of each Association in the U.S. and Canada in one regional meeting every other year. The plan was based on a recognized need to strengthen ties between the Federation and Member Associations by improving communications, allowing for the discussion of many matters of concern, and improving effectiveness of both Member Associations and the Federation.

The program was the outgrowth of an escalation in the Federation's programs and an increasing need for exchange of information which became evident during the late 1960's and early 1970's. An increased demand for discussions between Federation officers and Association officers became more and more evident. The busy nature of Association meetings and the involvement therein of the officers, who should benefit from the orientation sessions, hindered communication. Search for a way around this problem led to the plan for regional meetings, which officers and selected committee chairmen of various associations could attend to learn about Federation programs, discuss their own problems and solutions, and hear about the experiences of others.

Full support of Member Associations was necessary to ensure the success of these meetings, and attendance by key officials of the Associations was essential. In the beginning, it was suggested that the Association President, Vice President, Secretary-Treasurer, certain committee chairmen and other officers, and the Federation Directors should attend to meet with selected staff who would provide responses to questions. The meetings started with a dinner when the program for the following day was outlined, and leadership assignments were made. The next day covered the agenda agreed to during the evening meeting. The Federation hosted the dinner and luncheon and paid meeting room charges. Association representatives or their Associations paid individuals' travel costs.

The first regional meetings were held in Hartford, Atlanta, and Dallas during the spring and summer of 1973. These three meetings laid the basis for further development of the regional meeting concept and planning for Member Associations' officers' leadership and participation. The working dinner at Hartford on June 26, 1973 had 30 participants, including the Federation President and staff. The next day's meeting had 32 participants. Discussions included membership, publications, government affairs, education, operator certification, safety, public relations, and committees.

Based on the experience of 1973, plans for five additional regional meetings in 1974 were included in the budget. These included Washington, D.C. in January, San Francisco in February, Denver in February, Kansas City in April, and Chicago in August. Subjects discussed at these meetings varied somewhat but included most of the subjects discussed at the meetings of the previous year. The meetings showed that the communication afforded by the meetings was sorely needed. Association representatives had varied questions; many were easily resolved by furnishing the latest information available. Local recommendations at the meetings provided guidance for further Federation program development and the opportunity for encouraging greater participation on the part of the membership in the associations. Many questions were really complaints that had been raised in the past; some because new officers had not been briefed by outgoing officers, some because of inadequate explanations in the past. These meetings provided sufficient time and opportunity for discussion to resolve the questions and provide the needed explanations.

The early regional meetings were so successful that during the first two years numerous associations planned their own similar meetings in the two-year interval between the Federation-sponsored meetings. By 1976, the meetings had become so popular that a successful move was made during the Board's consideration of the budget for 1977 to have the Federation sponsor the meetings on an annual basis.

Publicity

A large amount of Federation publicity always centered on the Annual Conferences. At the Boston Conference in 1957, a press facility, established by the newly formed Public Relations Committee, provided to the working press, for the first time, telephones, typewriters, and news releases on conference events. This newsroom was used by both local media and national trade journals, and after 1957, operation of the newsroom was one of the main public relations activities during the conference. It was the backup for the main conference effort to inform the public about water pollution control. Almost 80 press representatives took advantage of the 1976 newsroom where news kits and conference abstracts were distributed. The growth in number of journalists using the newsroom reflects the success of the program as an informal way to feed news to and receive news from the press.

Probably the most important item distributed in the newsroom is the news kit. This practice began in 1960 as a "package news kit" for trade journalists, and by 1976 it was being distributed to all of the press covering the conference and other news outlets interested in conference events. The 1976 news kit was quite extensive. In addition to the news releases and abstracts, it contained biographies of Federation officers and key conference speakers, a membership brochure, a list of conference participants and their business affiliations, and a Statement of Policy.

In October 1969, the first pre-conference press luncheon was held to introduce the Federation and its new Executive Secretary, Robert A. Canham. News kits and a great amount of background material on the Federation were distributed. Later, follow-up letters and phone calls were used to determine what use the media was making of the information furnished at the luncheon.

Another publicity innovation, first tried in 1968 but greatly expanded in 1969 was the "self release"—it was a pre-written news release with blank spaces so that the conference registrant could fill in his name, employer, and sessions he would attend; he could then mail it to his hometown newspaper and/or other news media.

Successful conference publicity was also directed to authors and awards recipients. Advance copies of technical papers and news items were furnished prior to the conference for use by the press.

Boston's Mayor Kevin H. White proclaimed October 4-9, 1970, Water Pollution Control Federation week in that city. This honor was highlighted by a four page spread on the Federation in a special commemorative issue of the *Boston Herald-Traveler*.

As local and national news coverage of the annual conferences increased, *Engineering News Record* in 1971 began featuring the Federation and its president in a single issue. Of particular interest was the cover story on Joseph F. Lagnese, the new Federation President.

Pre-arranged interviews with Federation officers and news coverage of the 1971 conference produced a remarkable response with great area-wide media coverage which lasted from October 1 to 29. Also, as a result of this conference, Mr. Lagnese was featured in a *New York Times* "Man in the News" article.

The National Public Radio Network gave penetrating analyses of the Federation, the conference, and their environmental significance. Broadcasts by 102 affiliated stations included consecutive interviews from October 25 to October 29 throughout the country. Representatives of 35 publications helped reach many non-professional organizations. This distribution was significant because it provided added exposure to a conference that moves from city to city every year.

The pre-conference luncheon in 1972 tried to generate hard news stories for the press. This led to larger time allocations on radio and television and more expanded coverage.

Radio and television interviews with conference visitors from foreign countries were the publicity highlights of the 1973 conference. Most of the other Federation public relations activities at that time were strong follow-ups on the previous year's activities.

In 1974, the Monitor section of the *Journal* was sent to about 300 newspapers, trade and technical publications, editorial writers, environmental reporters, and the Washington, D.C. news outlets, including AP and UPI, in order to give the Federation's analysis of certain national issues.

A new display exhibit unit was first used and continuously manned at the 1975 WPCF annual conference so that registrants could speak with staff members on Federation activities. Also, postcards addressed to the Federation office in Washington, D.C. were available at the conference.

One of the most significant actions of the Public Relations Committee at the 1975 conference was completing the arrangements and working out the details for a new Public Relations Handbook. Only one other had been written by the Federation in its entire history, in 1965.

All through 1975, as in many previous years, important information programs were conducted by 30 Member Associations, including several overseas. These programs were mostly Federation literature displays consisting of promotional materials.

Due to a great amount of advance work, 1976 was a good year for publicity innovations at the conference. There were separate news releases on each Federation officer. Also of special interest was a newly created public relations display designed to promote the Federation and its activities in the water pollution control field.

An increasing demand for the text of program speakers by the press demonstrated the media's greatly increased interest in the water pollution control field.

There was a series of press "get-togethers," informal news conferences to help the press people familiarize themselves with prominent conference speakers and Federation officers and to learn what positions the Federation had or had not taken on water pollution control issues.

At the traditional pre-conference press luncheon in Minneapolis, President Wagner introduced the Federation to the Minneapolis press. There followed television and radio interviews with both President Wagner and Executive Secretary Canham, in addition to visits with editorial writers of two Minneapolis daily papers. The St. Paul papers printed the greatest number of stories on the conference and there was strong television and radio coverage of events and personalities.

An innovation of the 1976 public relations campaign was the convention-hall-wide broadcast of President Wagner's speech as he cut the ribbon to officially open the exhibition hall. The public relations efforts and activities had come a long way by 1976 from humble beginnings.

New Awards Established

One evidence of the Federation's continuing effort to recognize or promote water pollution control activity was the establishment in 1970 of the Harry Schlenz Medal. This medal honors one of the most energetic and loyal members and past presidents.

The Schlenz medal was intended to recognize people who are not in the pollution control field itself but who nevertheless make significant and outstanding contributions to pollution control. The award exemplifies Schlenz' idea that one must spend more time talking to people outside the technical field. He was always concerned with what other people thought of the Federation and its aims. Many times, he said "It's easy for one of us to contribute something to our own profession, but for others it is difficult."

The first presentation of the Schlenz medal was in 1971 to Gladwyn Hill of the New York *Times*, who through his writings and reporting provided distinguished service in promoting public understanding and action in water pollution control. Later awardees included Stuart Finley, Clem L. Rastatter, and Murray Stein.

The Member Association Safety Award was established in 1970 to stimulate Member Associations to promote vigorous safety programs in local wastewater works and to encourage the collection of injury statistics on a national basis. Important

WATER POLLUTION CONTROL FEDERATION

MEMBER ASSOCIATION SAFETY AWARD
PRESENTED TO

In recognition of the excellence of its program of promoting safety in water pollution control works including: service to its members; training, education and publicity on safety; cooperation with the Federation and other organizations to promote safety; and the encouragement of collection of injury statistics and other data to evaluate safety programs.

PRESIDENT   SECRETARY

WPCF's Member Association Safety Award initiated to stimulate vigorous safety programs.

The Collection System Award established in 1973 to recognize outstanding contributions made by individuals in advancing state of the art of wastewater collection.

Harry E. Schlenz for whom the Harry E. Schlenz Medal was established to recognize distinguished service in promoting public awareness, understanding, and action in water pollution control.

factors considered by the judges in making the award include: a Member Association's own safety program; cooperation with other organizations on matters of safety; safety publicity; materials and visual aids on safety developed by an Association; collection and use of injury data; and wastewater systems personnel injury experience during the previous five years in the area served by a Member Association. The first award was made in 1970 to the Indiana Water Pollution Control Association, with honorable mention to the Chesapeake Water Pollution Control Association. Other associations receiving the award later were California, New York, and Pennsylvania.

The WPCF Collection System Award was established in October 1973, and was first presented to Alvin A. Appel and Sidney Preen in 1974. The criteria used as the basis for the award are: (a) the nominee must have contributed, by original concept and outstanding practical application, to the advancement of the techniques of wastewater collection, and (b) the nominee's service must have been distinguished in any of the following areas: management, overall planning, operating and maintenance, facility design, education, training, or research. Other recipients include Horace L. Smith and Gerald D. Underwood.

Operators

For 50 years, improvement of the capability and status of wastewater works operators has been a prime Federation objective. In 1977, the Federation was still trying to help operators. The Federation always was concerned with and provided services to operators; indeed, it was originally the managers and operators who provided the impetus for the organization of the Federation. Those persons responsible for development, operation, and maintenance of collection and treatment facilities needed some organized publication of their problems and successes in order to share information on how to accomplish wastewater and pollution control. With an increasing number of facilities and a decreasing proportion of operating personnel requiring or appreciating the publications and services offered by the Federation, direction of the Federation program adjusted to others who did demand these services. Content of the *Journal* gradually became more technical and research oriented, or reported on the concerns of managing, evaluating, and planning for water pollution control.

Even with changes in the *Journal,* the Federation continued to provide special publications, training and certification materials, and other information which was helpful to operations personnel, even though the portion of the Federation membership represented by this group was small. On a number of occasions, efforts were made to increase operator memberships, which constituted 17 percent of Federation membership in 1969 and 14 percent in 1976.

In 1969, the operator training workshop at the Annual Conference in Dallas made a special effort to evaluate the problems of operator recruitment, training, certification, and retention. The outgrowth of this workshop was a strong recommendation, approved by the Federation Board of Control, to provide leadership in this area and solicit the cooperation of others in developing a national program.

An attempt at initiating this program was made under the title of MANFORCE (MANpower FOR a Clean Environment) (Chapter VI). At the same time, forces outside the Federation were moving in a similar direction with strong financial support. A variety of manpower development and training programs were started or expanded. The activity was very popular but most of the national financial support came from sources having the objective of increasing employment of minorities and others in low-income groups. As a result, the addition of skilled and permanent operators was more the exception than the rule.

Although the Federation MANFORCE program was short-lived in name and original concept, it realized several lasting achievements and improved the Federation's ability to assist operating personnel. During this time, *Deeds & Data* was first printed, and the "MTM-01 Extended Aeration Process Control" training materials were developed.

Another Federation effort toward operator improvement or provision of services was the safety program. Assistance in safety promotion, training materials, and special seminars on safety were provided through Federation Member Associations and municipal and other agencies, with an annual survey to determine the extent of injuries in wastewater works. Special efforts to obtain and publish information about safety problems over a period of ten years saw the injury rate decrease and then increase. In 1976, two Safety Committee members and one staff member assisted in organizing and became members of a new National Safety Council division on wastewater in the Public Employee Section of the Council. The purpose of the new division was to expand wastewater works safety programs.

During the early 1970's, special studies on how to promote operator membership resulted in a decision to promote a member association classification for operators which did not require membership in the Federation. At the Membership Associations' option, such a classification would provide for a special low cost subscription to *Highlights/Deeds & Data*. This plan was developed and included in a revision of the Federation's Bylaws. Many Member Associations incorporated this "affiliate" classification with the result that operator membership began to increase in Member Associations.

One of the factors that had inhibited operator membership in the Federation undoubtedly was the salary level of operators. Despite the Federation's concern about this situation, previous salary studies generally had been limited to state and regional areas. Early in 1977, the Personnel Advancement Committee, with staff assistance, conducted a national survey of operators' salaries and fringe benefits.

Summary

The Federation's original objective was to provide a means of publishing material for the fledging water pollution control industry. Then came the annual conferences, expansion of the publication program, and other services to Federation members, to the industry, and to the public. At the end of 50 years, however, the publications area still represented nearly 50 percent of the total effort of the Federation.

For most of its first 50 years the Federation moved along a steady but highly conservative path quietly doing the job of serving its members. This was particularly true before the federal government became so heavily involved in pollution control.

After 1956, with the enactment of a federal law which included subsidies for construction of water pollution control facilities, a few farsighted Federation leaders predicted a different kind of relationship between the water pollution control technical community and regulatory authorities. This change in relationship and its associated problems, accelerated with each new federal law, culminated with the passage of PL 92-500. With so many more employees in the Federal program, who were mostly new to the water pollution control field and who were more closely directed by political appointees, problems of communication between the Federation and the government were compounded.

The Federation found it necessary to make adjustments and began to make them. While the Federation remained a technical organization, there developed an increasing number of reasons for it to change from a passive to a more activist role. This change was not easy and was still being developed in 1976-77.

Over the years, the Federation made significant progress in responding to the needs of its members and in filling its proper role of providing substantial and constructive input to the national program. At the end of its 50th year, it looked forward to greater expansion, greater service, and greater effectiveness of effort. Its integrity was assured, its position accepted, and its recognition an accomplished fact.

Presidents and Executive Secretary, 1970-77

J. B. Hanlon
1969-70

A. F. Vondrick
1970-71

J. F. Lagnese, Jr.
1971-72

J. F. Byrd
1972-73

J. D. Parkhurst
1973-74

S. L. Warrington
1974-75

V. G. Wagner
1975-76

H. L. Smith
1976-77

R. A. Canham
1969-77

Chapter Six

Federation Programs

As pointed out in Chapter I, the need for a journal in which those interested in wastewater could publish was the primary reason for organizing the Federation, and its *Sewage Works Journal* was the Federation's first program. In 1940, the series of annual conferences was inaugurated; in the years that followed, more programs were added: education, governmental affairs, Manuals of Practice, and technical services. The activities in these areas are discussed in the following pages.

The Journal

According to C.A. Emerson's review of the first quarter century of Federation history,

> The first number of *Sewage Works Journal* appeared late in October 1928, and was sent to 273 subscribers.
>
> From the start there had always been more than sufficient excellent material pressing for publication, with result that the volume for 1931 contained 793 pages of editorial matter, or almost double the number initially contemplated. It was then decided to issue the *Journal* bi-monthly, instead of quarterly, but without increase in dues. This was an unusual step in the midst of the "Great Depression," but as there was about $2,000 in the treasury, the chance was taken.
>
> In 1937, the development of Federation activities and rising costs in publication necessitated an increase in subscription price to $1.50. Thereafter, the continued rise in printing costs and in volume of editorial matter necessitated further increases to the the present (1952) price of $5.00 for Active Members, as instituted in January 1950, when the *Journal* became a monthly publication. The Federation has never made any money on the Active Member subscription price for the *Journal*. From the very start, the actual cost of printing and mailing, exclusive of editorial and overhead expenses, has equalled or exceeded the subscription price.
>
> The *Journal* has kept fully abreast of developments in the art of sewage treatment. Its outstanding success in the United States and abroad has, of course, been due in major part to the untiring efforts of Dr. Mohlman, who bore the burden of editorship for 13 years and continues to this day as Advisory Editor.

An insight into the early Federation years appears in the following letter:

Nov. 12th, 1928
Taylor & Woltman

Gentlemen:

Referring to our letter dated August 7th, we are enclosing herewith a copy of the first issue of Sewage Works Journal for your inspection.

The Journal will appear quarterly and each issue will contain about 100 pages of reading matter. An advertising contract and rate card are enclosed for your convenience. The next number will go to press on December 1st, and we would suggest that you send in your advertising contract together with your copy for the January number as soon as possible.

Yours very truly,

Wm. W. Buffum, Business Manager

With the letter from Mr. Buffum was enclosed the advertising contract which is reproduced in the Appendix.

In the first 13 years of the Federation, the primary objective had been to produce a viable technical publication to further the wide range of science and technology applicable to the prevention and abatement of water pollution. By 1940, *Sewage Works Journal* was technically excellent, but its financial base was tenuous at best. There were no resources to support the other membership and public services that were envisioned by the Committee on Expansion and Reorganization in 1940.

When the Federation was reorganized, the change in the Federation and the *Journal* was explained to the Association Secretaries in a letter from Emerson, reproduced in its entirety in the Appendix.

Emerson said, in 1952, "Although the Federation probably could have existed without the backing of Chemical Foundation, it is certain that the approximately $6,750 cash they contributed, plus the efficient and painstaking efforts of their staff in management of the *Journal,* contributed greatly to the rapid growth of the Federation in its early years."

The part that the Chemical Foundation had in the publication of the *Journal* was appropriately acknowledged by an editorial in the March 1944 *Journal.* Major excerpts of that editorial also appear in the Appendix.

Evolution of the Journal

By 1941, *Sewage Works Journal* had established itself under Dr. Mohlman as a first-rate scientific/engineering publication. Issued quarterly from October 1928 to 1931, then bimonthly and eventually monthly, it carried the best material available on research, design, operation, and management of waterborne waste collection and treatment works, and on sanitation of natural waters. In only 13 years, the *Journal* had acquired international recognition.

The fiscal position of the *Journal*, however, was precarious. In 1940, income from subscriptions and advertising exceeded production costs by only $768. Advertising netted only $6,000 after commissions to the two space salesmen then under contract.

Although totally inexperienced in the advertising field, Executive Secretary Wisely, in his capacity as managing editor of the *Journal*, gave high priority in 1941 and thereafter to increasing the revenue from this source. A space sales brochure was prepared, and a continuing mail promotion campaign was initiated. A special Annual Convention issue was inaugurated in 1941, bringing in 18 pages of new space that year. The two space salesmen, who did little to earn their commissions, were released with the expiration of their contracts.

Advertising income doubled from 1940 to 1944 and then almost quadrupled in the next 10 years. By 1954, annual revenue from advertising was $49,123, comprising a third of the Federation's total operating income.

A primary objective of the 1941 expansion and reorganization program was to increase the value of the Federation, and the *Journal* in particular, to the sewage works operator. It was difficult, however, to achieve an editorial balance that would serve the needs of the professional engineer and scientist on the one hand and the sewage works technician on the other.

Editor Mohlman and Executive Secretary Wisely resolved this problem by assignment of a portion of each issue of the *Journal* to be devoted to practical, workaday material, presented in a form useful to operators of large and small plants. Responsibility for "The Operator's Corner" was assumed by the

Executive Secretary, who, at the time, was spending half of his time in sewage works management. One feature of the new section was the column "Bark from the Daily Log," which summarized actual operating experiences that were informative and of general interest. More than 250 pages were given over to "The Operator's Corner" in each of the years from 1941 to 1944. The "Daily Log" column was discontinued in 1944 when the Executive Secretary gave up his work with the sanitary district to devote full time to the Federation, but "The Operator's Corner" was a continuous feature of the *Journal* until 1961.

The volume of material meriting *Journal* publication had grown by 1949 to such an extent that monthly publication was given study. A 1950 decision to give specific emphasis to industrial wastes management in the program and name of the Federation resulted in changing the name of the publication to *Journal of Sewage and Industrial Wastes;* a monthly publication schedule was adopted.

By 1950, the *Journal* had become such a significant scientific and technical reference that there was wide demand for an index of past issues. A 20-year index (1928-48) was compiled and published, to be followed later by one 10-year and a series of 5-year cumulative indexes. A system of key words was introduced that was unconventional at the time, but which has since become a standard indexing procedure.

The Federation's *Journal* was firmly established by 1950, and the switch to a monthly publication schedule provided a timely medium for the interchange of technical information and ideas. During the decade of the 1950's, the *Journal* consolidated its position as the leading monthly publication in its field, and with the granting of Member Association status to non-U.S. organizations, the readership of the *Journal* began to extend throughout the world. Not only was the editorial page budget of the *Journal* increased, but the scope of its technical content gradually expanded in response to the expectations of both its readers and contributors. The increased emphasis on industrial waste control that lay behind the change of title in 1950 was reflected in the editorial pages; papers on process research and design became more numerous, as well as studies on stream surveys and effects of pollutants on natural waters.

Covers of three *Journals,* all the same except that titles were changed in keeping with the name changes of the organization.

Late in the 1950's, after passage of the 1956 Water Pollution Control Act, the *Journal* began what was to become a frequent practice of commenting on the efficacy of the legislation affecting the national water pollution control program.

The trend toward a broadened editorial scope of the *Journal* that began in the 1950's continued into the next decade. By 1960, the number of editorial pages (exclusive of the Yearbook/ Directory and advertising pages) had reached 1,384; this grew to 1,806 in 1965, 2,222 in 1970, and to 2,916 in 1975.

Annual Literature Review

The publication of technical papers was not the *Journal's* only function, and a number of monthly features and special issues were introduced in the 1950's that were received by Federation members. Most notable of these was the *Journal's* Annual Literature Review. Its growth, particularly in the later years of the period, was almost as great as its acceptance among *Journal* readers.

Reviewing the technical literature of the water pollution control field became a tradition with the *Journal* almost from its inception. From the very beginning, one *Journal* issue each year devoted a portion of its pages to a summary of the developments of the previous year as reflected by what was being published in other technical journals. The basis of such reviews was that communication among engineers and scientists in a wide range of disciplines is fundamental to progress in the field of environmental improvement. Without such communication, the potential for progress in a field as broad as water pollution control would be severely limited.

It was not until the early 1930's, however, when the Research Committee was formed, that the Annual Literature Review took shape as a formal activity within the Federation. From then on the scope of the Reviews grew steadily as the Committee's size increased to allow existing review categories to be expanded and new ones added. By 1940, the *Journal* published the Review each May; by 1950 the size of the Reviews had grown so that it was necessary to publish them as parts of the May and June issues. In 1953, the Reviews required portions of the May, June and July issues.

The largest growth in the size of the Reviews occurred in the decade of the 1960's, when the number of *Journal* pages devoted to them more than tripled. Beginning in 1968, the Annual Literature Review became one entire issue published in June. After 1968, the Reviews continued to grow; in fact, in recent years the Research Committee faced the formidable challenge of controlling the size of the reviews, for economic reasons, without sacrificing their quality or thoroughness.

As with so many Federation programs, the Annual Literature Review would not be possible without the efforts of a volunteer committee. Long before the Federation's 50th year, the WPCF Research Committee had become one of the largest of the standing committees of the Federation. Each year, this Committee devoted considerable time and effort to compiling, preparing, and editing material for the Review issue.

A cursory look at any recent issue of the Review reveals the wide range of talents this Committee brings to its task. By 1976, the Review covered many disciplines, including several branches of engineering, biology, chemistry, and public health, among others. Although often considered solely a research publication, the Review contains material on process development, facilities design, economics, operations, and administration. It is this attempt at both breadth and depth that produced the Review's wide acceptance and growth.

In a readership survey conducted in 1975, 79.7 percent of the respondents indicated that they found the Annual Literature Review issue of the *Journal* directly useful in their work. Another 12.8 percent found no direct use for the Reviews but believed the feature was a worthwhile service that should be continued as a special *Journal* issue. This 92 percent vote of confidence was a tribute to all past members of the Research Committee who were active contributors to the Reviews.

Research Supplement

Another innovative *Journal* feature, a series of separate research supplements, did not prove as successful as the Annual Literature Review. Early in 1967 federal grant assistance became available through the Federal Water Pollution Control Administration that could be used to partially subsidize the publication of research results in the *Journal*. The FWPCA apparently felt that such a grant would be one way to speed up the dissemination of research data urgently needed for its rapidly growing pollution abatement program. From the Federation's point of view, the grant possibility offered a chance to evaluate the economic feasibility of publishing a separate research journal, one that might eventually become self-sustaining through a combination of advertising support and member subscriptions.

The grant was applied for and obtained, and in October, 1967, the Federation undertook the publication of a series of quarterly "Research Supplements" to the *Journal*. These supplements were prepared separately from the regular *Journal* issues, but were mailed to all *Journal* recipients every three months along with the regular issues for those months. Twelve issues were eventually published.

During the three years of publication, the Research Supplements did serve a vital purpose. There can be little doubt about their acceptance by *Journal* readers. Furthermore, the federal

Cover of one of the Research Supplements, which were published in addition to the Journal during the period of October 1967 through August 1970.

grant support allowed the Journal to speed up publication of a growing backlog of research papers that had been accepted for publication but, because of the relatively limited Journal editorial page budget, had to wait as long as 18 months for publication. In the later stages of the project, however, it became obvious that economic self-sufficiency for the Research Supplements would take considerable time to develop. Some advertising support for their publication was generated, but not enough to significantly offset publication costs; and because of the limited circulation projected if the Supplements were sold separately, the subscription costs would have been prohibitively high.

At the same time, federal interest for the project began to wane; grant support was in fact cut back on the second and third years of the project. With the continued outside support for the Supplements that was needed to ensure their viability becoming increasingly uncertain, the decision was made to discontinue their publication with the August, 1970 issue.

Monitor Section

The latest innovation in the Journal was the development of its Monitor section, devoted to interpretive reporting of current events in the water pollution control field. The initial development of this concept originated in the Publications Committee, which in 1971 was seeking ways to enhance the Journal by presenting material of interest to all segments of its widely diversified readership. The eventual intent was to devote a number of pages each month to coverage of the water pollution control field in a news context, and to report on such things as changes in environmental policy in the U.S. and abroad, accounts of the development and implementation of new technology, and interviews with leading spokesmen and opinion makers in the water pollution control field.

In consultation with the Federation staff, the Publications Committee developed a detailed proposal for publication of such a *Journal* feature and presented it to the Board of Control in October 1971, where it generated considerable discussion. Opinion on the proposal was about equally divided between those in favor of it and those who, though not necessarily opposed to the idea in principle, had some reservations about the budget increase and some misgivings about the implementation of the concept. After a vigorous debate, the proposal was defeated by two votes. Immediately following the vote, the Board directed the Publication Committee and staff to refine the proposal, particularly its budget impact, and resubmit it to the Board. The new proposal was passed by a large majority in October 1972, and publication of the news section began in 1973.

The Monitor section was gradually expanded in scope and content and its acceptance was generally favorable among *Journal* readers, who quickly perceived it as a means of making the *Journal* more reflective of current events in the pollution control field. This attitude led to the *Journal's* being perceived by its advertisers as a more viable and desirable advertising medium; in fact, the introduction of the Monitor section had a demonstrable impact on advertising revenues. Not only did the overall number of paid ad pages increase (in a period of declining revenues for most other technical magazines and journals) but more and more advertisers began using two- and four-color display advertisements. This increase in *Journal* advertising revenues, along with the installation by the *Journal* printer of high-speed, web offset printing equipment, made possible some gradual refinements in *Journal* format. Among these were the more frequent use of color in the editorial section of the *Joural,* particularly in special issues and features; a gradual modernization in typography; and, most noticably, the use of graphic designs and four-color art work on the cover of the *Journal*.

Thus, the *Journal* continued its gradual evolution as the leading publication in the water pollution control field. It became the preferred medium of publication for the most knowledgeable and proficient authors for presentation of their data, and, as a result, the advertising revenues it generated contributed significantly to its economic stability.

Other Journal Innovations

Along with the growth of international interests, the practice of including multilingual abstracts in the *Journal* was initiated in 1966. Thereafter the *Journal* included monthly translations of abstracts of its articles in French, German, Spanish, and Portuguese. This practice was intended to make the *Journal* more useful to people in those countries. In 1963, the *Journal* began using metric terms in parenthesis after English measurements in order to educate its U.S. readers in the use of the metric system. This feature also appealed to many overseas readers. In 1976, SI or Systeme International d'Unites was adapted as the primary system of units for the *Journal*.

In 1966, the *Journal* adopted the feature of including information retrieval forms for indexing of articles by the reader. This feature followed the form suggested by the Engineer's Joint Council.

Highlights

Although the *Journal* had always been and continues to be the major publication, it was only one aspect of the overall WPCF publications and communications effort. In the early 1960's, the Federation discovered that although the *Journal* had become well established and was performing a vital service, more than one publication was necessary for a growing organization to serve its members effectively. Accordingly, at its October 10, 1963 meeting in Seattle, the Board of Control authorized the publication of a monthly newsletter titled *Water Pollution Control Federation Highlights*. Its purpose was to bring news of the Federation and the water pollution control field to the prompt attention of its members.

The Federation newsletter *Highlights* was begun as a monthly periodical in January 1963 and its supplement, *Deeds & Data*, commenced publication in June 1970.

The newsletter, as stated in its inaugural edition in January 1964, was "to be a monthly publication designed to provide timely and interesting material. It will be kept brief to allow rapid reading, and will attempt to serve the full range of Federation membership. To do this it will include such things as specific news items; organizational information; comments on the field such as legislation, operation, certification, management, safety, and other areas; public relations tips; and selected short papers of general interest."

The January 1964 issue of *Highlights* was nominally designated Volume 1, Number 1; strictly speaking, it was not the first issue. In November 1963, a month after the Board had authorized the publication of *Highlights,* a special issue was prepared; it contained the full text of a luncheon address by John Charles Daly that had caused a stir among the audience at the WPCF Seattle Conference. The subject of Mr. Daly's talk was the growing sentiment for removing the national water pollution control program from the U.S. Public Health Service, a move which both he and the Federation had vigorously opposed.

Other items from the first year of publication, apart from their historical value, give some flavor of the early content of the publication and an idea of how soon its editorial goals were achieved. In that year, items in *Highlights:* described the ceremony honoring the 10,000th member of the Federation (William C. Henry, of Wilmington, Del.); announced the move of the Federation offices to a new location because of additional space requirements; informed the membership of the establishment of the Federation Life Member Award; and pointed out that, in 1963, wastewater construction contracts amounted to a record-setting $1.03 billion. The tone and content of those initial issues, with few exceptions and only minor increases in page budget, were maintained thereafter.

Deeds and Data

Another significant event in the development of the WPCF publications program occurred in 1970. A survey conducted in that year by the Membership Liaison Committee revealed an urgent need for a substantial increase in services to the operator members of the Federation. In addition, several of the WPCF Member Associations had passed resolutions requesting more attention to operators' needs and stimulation of increased participation by operators in Federation affairs. In June 1970, the Federation responded to this need and began publishing *Deeds & Data,* designed especially for operation, maintenance, and management personnel in water pollution control facilities. The goals of the publication were stated by WPCF President Joseph Hanlon in its first edition:

> Effective operation and maintenance of all new wastewater treatment plants are now prerequisites for federal grant approval. Also, the need to improve and upgrade the operating standards of all existing installations is demanded. We at the Federation accept this challenge and look forward to contributing to this effort with greatly expanded services to our operator membership and the entire water pollution control industry.

Deeds & Data was originally conceived as a four-page bimonthly publication to be prepared along with and mailed as part of *Highlights*. Its acceptance was immediately enthusiastic, and soon generated pressures for expansion of the effort. In January 1971, just six months after its introduction, *Deeds & Data* began appearing every month, and before the year was out its content was increased to eight pages.

In 1976, its monthly page budget varied between 8 and 12 pages, and in addition to short technical papers on operation and maintenance of treatment facilities and collection systems, it contained several popular monthly features and services such as an employment service, a certification quiz, letters from operators, and schedules for operator training courses.

Manuals of Practice

Another recommendation of the 1941 Committee on Expansion and Reorganization was the initiation of a series of manuals of sewage works practice, to provide authoritative guidance in specific problem areas. It was envisioned that such manuals would eventually be assembled into a single volume.

The first manual, "Occupational Hazards in the Operation of Sewage Works," was published in 1944, authored by an able committee under the chairmanship of LeRoy W. Van Kleeck. The immediate success of the "Safety Manual" ensured the future of the entire program. The second manual, "Utilization of Sewage Sludge as Fertilizer," followed in 1946; No. 3 on "Municipal Sewer Ordinances" and a tentative manual "Uniform System of Accounts for Sewer Utilities," in 1949; No. 4 on "Chlorination of Sewage and Industrial Wastes" in 1951; and No. 5 on "Air Diffusion in Sewage Works" in 1952. Manuals on units of expression, protective coatings, storm and sanitary sewer design, treatment plant design, operation of treatment plants, and sewer maintenance were started by the mid 1950's.

The Manuals of Practice have had significant impact on the water pollution control field. Each became a primary reference work in its particular subject area. For example, MOP's 8 and 9, both published jointly with the American Society of Civil Engineers, comprise the most complete works on wastewater collection and treatment facilities design in this country, and have traditionally been among the largest sellers in the Manuals of Practice series; in 1976 they were very much in demand among designers, state control agencies, and sanitary engineering departments of major universities. Likewise, Manual of Practice 11, "Operation of Wastewater Treatment Plants," became the standard work on the subject soon after its issuance in 1961. This 150-page work was substantially updated and considerably

TABLE VI-1. WPCF Manuals of Practice

No.	Title	1st Edition	Current Edition
1	Safety in Wastewater Works (Original title: Occupational Hazards in the Operation of Sewage Works)	1944	1975
2	Utilization of Municipal Wastewater Sludge	1946	1971
3	Regulation of Sewer Use (Original Title: Municipal Sewer Ordinances)	1949	1975
4	Chlorination of Wastewater	1951	1976
5	Aeration in Wastewater Treatment	1952	1971
6	Units of Expression for Wastewater Treatment	1958	1976
7	Sewer Maintenance	1960	1966
8	Wastewater Treatment Plant Design (Joint with ASCE)	1959	1977
9	Design and Construction of Sanitary and Storm Sewers (Joint with ASCE)	1960	1969
10	Uniform System of Accounts for Wastewater Utilities	1949	1970
11	Operation of Wastewater Treatment Plants	1961	1976
16	Anaerobic Sludge Digestion	1968	1968
17	Paints and Protective Coatings for Wastewater Treatment Facilities	1969	1969
20	Sludge Dewatering	1969	1969

Note: Publication Nos. 12, 13, 14, 15, and 19 are not MOP's; see Table VI-2.

expanded to more than 500 pages in 1976. The reception of the 1976 edition which includes color plates (a first for WPCF manuals) appeared to promise a real improvement over the response to the previous edition.

The development of these Manuals of Practice demonstrated a sense of professional dedication among Federation members and their willingness to work through the established committee structure towards a common objective. All of the MOP's were produced under the direction of the WPCF Technical Practice Comittee, which consists of a carefully selected group of experts in their various fields.

The Manuals of Practice series was a successful financial venture from the start and an eminently successful venture technically. Publication of the first few manuals in the series was financed by a revolving fund of $5,000 appropriated in the 1944 budget; ten years later the subsidy had been reduced to $3,000 and the fund enjoyed a balance of $8,300 plus a saleable inventory of more than $2,000. The need for such a revolving subsidy fund disappeared not long after, and the Manuals of Practice stood alone as a self-supporting venture. For example, in 1976, the Manuals produced a gross income to the Federation of more than $150,000 and in the first four months of 1977 more than $100,000.

Joint Publications

In addition to its own publications, the Federation collaborated with other technical and professional organizations in several noteworthy publishing efforts. The joint effort with the American Society of Civil Engineers for the production of design books on sewers (MOP 9) and treatment works (MOP 8) was noted previously.

In 1935, the Federation provided a report on standard methods of sewage analysis which was incorporated in the eighth edition of "Standard Methods for the Examination of Water and Sewage." Previously, coverage in "Standard Methods" had been limited to water analysis. The same contribution was made to the ninth edition in 1946, and a year later the Federation became a full partner with AWWA and APHA in the produc-

TABLE VI-2. Other WPCF Publiactions

Title	Year issued
Public Relations for Water Pollution Control (Publ. No. 12)	First issue in 1965; revised edition to be released in 1977.
Simplified Laboratory Procedures for Wastewater Examination (Publ. No. 18)	First issued in 1968 with several reprintings; revised edition published in 1976.
Well of the World (Pub. No. 19)	Published in 1967; currently out of print.
Standard Methods for the Examination of Water and Wastewater	First edition published in 1918 by APHA; WPCF became co-publisher of 10th edition in 1955; current edition (14th) published 1976 by APHA, AWWA, WPCF.
Financing and Charges for Wastewater Systems	Published in 1973 in conjunction with APWA, ASCE.
Joint Treatment of Industrial and Municipal Wastewater	Published 1976.
Glossary—Water and Sewage Control Engineering	First published in 1949 with ASCE, AWWA, and APHA; revised in 1969.

tion of the 10th edition (1955) of this important laboratory manual.

The "Glossary—Water and Sewage Control Engineering," published in 1949, was a joint venture of the Federation with ASCE, AWWA, and APHA. This manual was revised in 1969 by organizational committees under a joint editorial board. In 1976 a new joint editorial board was organized, with WPCF's representative Bob Rogers serving as chairman, to again revise this manual to meet the needs of the rapidly expanding and increasingly diversified field.

Modern Sewage Disposal

On the occasion of the tenth anniversary of the Federation's founding, an anniversary book was published. Titled "Modern Sewage Disposal" and edited by Langdon Pearse, the dedication of the flyleaf reads:

> This Anniversary Volume is issued to commemorate
> TEN YEARS OF SERVICE
> by the
> FEDERATION OF SEWAGE WORKS ASSOCIATIONS
> and is dedicated to the many
> OPERATORS OF SEWAGE WORKS,
> RESEARCH WORKERS, ENGINEERS, and
> PUBLIC OFFICIALS
> who have contributed to the development of
> the art of sewage treatment
> in the past

Opposite the first page of the foreword by C.A. Emerson, Jr., was a simple "In Memoriam" to Kenneth Allen, Harrison P. Eddy, George W. Fuller, Francis P. Garvin, John H. Gregory.

These names were still revered in the 1970's for their contribution to both the technology of water pollution control and the founding of the Federation.

Emerson's foreword to "Modern Sewage Disposal" recites briefly the story told in Chapter I of this history. Charles Gilman Hyde, Professor of Sanitary Engineering at the University of California, Berkeley, wrote an introduction titled "A Review of Progress in Sewage Treatment During the Past Fifty Years

A 371-page volume, "Modern Sewage Disposal," was published to commemorate the tenth anniversary of the Federation. Publication was made possible by editorial efforts of Central States member Langdon Pearse.

in the United States." The entire table of contents of the volume is printed in the Appendix to show the breadth of the technology just ten years after the Federation came into being.

The list of authors of the 1938 publication includes most of the top names in the then new industry of water pollution control. Now out of print, the book consisted of 371 pages containing, as of that time, the latest facts and theories of the various processes used for sewage treatment. The anniversary volume reveals the growth of municipal sewage treatment from 1855 on, and would be interesting to engineers or to anyone writing a text book on the history of environmental engineering. Among its highlights is a chapter on "The Works Laboratory" by W.D. Hatfield, who 20 years later was Federation President, 1958-59. Most of the book was devoted to treatment plant operation, with smaller sections covering research, regional and national aspects and industrial wastes.

WPCF Annual Conferences

From 1928 through 1939, annual meetings of the Water Pollution Control Federation consisted of meetings of the Board of Control. The 1928 Board meeting was in Chicago and from 1929 through 1939 Board meetings were held in New York. Beginning in 1940 (October 3-5), Annual Conferences of members were held each year, with the exception of 1945. The conference attendance in 1941 in New York was 556; the 49th

Annual Conference held in Minneapolis in 1976 was attended by 8,716 persons.

From 1940 through 1947, Annual Conferences were 3-day meetings of single sessions. In 1941, the Board of Control held its reorganization meeting in January, and the second national general convention for the membership was held in October, both in New York City. Beginning in 1942, the business sessions of the Board were held at the time of the annual October conferences. This arrangement was in accord with the expansion plan.

In 1948, the conference was expanded to 4-day meetings and in 1956 concurrent Technical Sessions were introduced. Beginning in 1960 the Conference was further expanded to a 5-day meeting with the first day being devoted to Board of Control and Committee meetings. The 1940 Conference consisted of five sessions and 13 technical papers. The numbers grew over the years to 37 sessions and 146 technical papers at the 1976 meeting.

Cutting the ribbon to the 49th Annual WPCF Conference and Exhibits.

Conference sites were selected from invitations issued by local Member Associations on behalf of cities in their geographical area, with the Conference Site Committee recommending their final selection to the Board of Control. By the early 1970's, Conference sites were being selected eight years in advance. An effort was always made to maintain a geographical balance in the selection of sites, but by the end of the Federation's fourth decade, the size of the Conference limited the number of cities that could be selected. Information on the conferences and the attendance records through 1976 are shown in the Appendix.

Another valuable part of the annual meeting was the manufacturers' exhibits. In 1940, the Water and Sewage Works Manufacturers Associations (WSWMA) gave the Federation $1,500 to help cover convention expenses and offered to contribute $5,000 annually to the general fund of the Federation in return for "the privilege of taking charge of the exhibits at

Federation exhibits at the 49th Annual Conference, Minneapolis, Minn., October 3-8, 1976.

the Annual Meetings . . ." The offer was promptly accepted, and the arrangement was finalized in 1941. Not only did the WSWMA (later re-named WWEMA) assume full responsibility for producing the exhibit, but it also established operating rules which allowed for smoother scheduling with the technical sessions. The WSWMA contribution was increased from $5,000 to $6,000 in 1949 and to $7,500 in 1960. From 1965 through 1967, contributions from WWEMA to the Federation increased steadily, and reached $93,000 in 1972; however, throughout this period there existed no contractual agreement between WWEMA and WPCF, nor did the Federation have any control over the management of the exhibits program. In the early 1970's, as a result of extensive negotiations with WWEMA, an agreement was reached whereby WWEMA would continue management of the exhibits, with WPCF participation, until 1973. Thereafter, the Federation assumed full control over the management of the conference exhibits.

Early Meetings

Many of the early annual meetings were memorable ones. In an article "Sewage Works Federation Throws Off Swaddling Clothes," *Water Works & Sewerage* devoted over 12 pages in two issues to the third annual conference held in Chicago in October 1943. The following quotation is from the first two paragraphs of that report:

> To have attended the Chicago Conference of the three-year-old Sewage Works Federation on October 21-22-23 was to be impressed with the healthy progress made by this young national organization, which was born in Chicago and returned there just three years later, to hold what is generally considered its best annual meeting from all points of view.

> To begin with, the attendance was the highest recorded for the four meetings. With a paid registration of 614, the attendance is estimated at 700 or better. The more impressive does this new record become when considering the times, travel difficulties, and discouragement of would-be attendants in finding hotels sold out many months in advance of the conference dates.

Highlights of the Chicago meeting included the completion of George J. Schroepfer's term as the third president and the induction of A M Rawn, plus addresses by Mayor Edward J,

Kelly of Chicago and Maury Maverick, Director of the Government Division of the War Production Board. Charles Gilman Hyde and H.E. Moses were elected honorary members. Entertainment was arranged by Harry Schlenz (Federation President in 1961-62). Thirty-four firms were represented in the exhibit hall. The program committee chairman was F.W. Gilcreas.

At the 1944 meeting, a strike at the headquarters hotel in Pittsburgh was settled only a few days before the convention (a cancellation notice was already drafted for release to the membership if it became necessary). That meeting featured a demonstration of safety equipment; the attendance trophy was won for the fourth time by the Central States association; attendance was 523, including 35 spouses, and there were 31 manufacturers' exhibits.

The 1945 meeting was cancelled because of the war. A Board of Control meeting was held, however, in Chicago. Several associations also gave up meetings that year at the government's request.

In 1946, at Toronto, Ontario a new attendance record was set, but the number of exhibits (35) was disappointing. That the meeting itself was not a disappointment is attested to by these paragraphs from *Water & Sewage Works:*

> When one begins to write about the 19th Annual Meeting of the Federation of Sewage Works Associations at Toronto on October 7-9, it is difficult to find sufficient superlatives to adequately express the proper appreciation and appraisal of this, the largest meeting of the Federation yet held.
>
> While the official registration of 812 topped the previous high by an even 200, it is known that there were at least another two dozen or so persons who attended some phase of the convention but did not register. In addition to setting a new high registration at which future meetings must shoot, this meeting set many other unusual precedents of interest.
>
> For one thing, the gracious hospitality of the Canadians and the Canadian Institute on Sewage and Sanitation as hosts is something to remember with delight. The experiment of beginning the convention on Monday (board meeting on Sunday) and ending on Wednesday seemed to meet with general approval. A chance to participate in the guided discussion type of operators' forum as it is carried on in Canada repaid many persons for the trip.

> . . . the quietly efficient organization and operation of the mechanics of the convention [were] a tribute to the ubiquitous spark plug, Dr. A.E. Berry, Secretary of the Canadian Association. [Dr. Berry was Federation President in 1944-45.]
>
> Although the number of exhibitors was relatively small (35), the exhibits themselves were interesting and informative, several new pieces of equipment being unveiled for the first time.
>
> A reporter of this meeting would be distinctly remiss in his duty if, in his enthusiastic praise, he failed to mention the work of Secretary W.H. Wisely and the officers of the Federation who presided and otherwise aided in the mechanics of the meeting. And having thus expressed an opinion shared by many, this reporter can now proceed with the more prosaic story of the meeting.

The "prosaic" part of the meeting included a "sludge disposal symposium," a session on industrial waste treatment featuring the General Electric Co. film titled "Clean Waters," a session on sewage systems and treatment practices, and the popular "Operator's Forum." At that conference, "Bill" Orchard was presented honorary membership with this citation:

> For his early guidance of the Federation in financial matters . . . his individual efforts in securing sufficient advertising pledges to insure success of the original quarterly journal; his service on the original Committee of One Hundred and the Temporary Committee which launched the Federation in 1928; his 16 years of faithful service on the Board of Control as Director-at-Large, and his continuing active participation in the administration of the Federation as Chairman of the Finance Advisory Committee.

Regarding the 1947 Joint Meeting with AWWA in San Francisco, Secretary Wisely wrote in retrospect, that it "was a gratifying exercise in inter-association cooperation, successful beyond all expectations. The courtesy and helpfulness of AWWA Secretary, Harry E. Jordan, was unbounded on this occasion." The reportorial account of the 1947 meeting by the associate editor of *Water & Sewage Works* contained these opening paragraphs:

> With the City of the Golden Gate offering visitors a sample of its very finest weather, the week of July 21-25 saw water and sewage works men and their ladies gathered from all over the country in San Francisco, Calif., for the first concurrent convention ever held of the American Water Works Association and the

Federation of Sewage Works Associations. Registration figures indicated the record combined attendance of 1955, of which 1,426 were men and 529 their ladies. (Federation member registration was 964). Manufacturers' representatives accounted for but 401 of the 1955 gross attendance, and except for about 600 from California, the remainder (1,355) came from various points in the country—many on special trains from the East and Midwest.

Further statistics indicate that there were 60 papers and one motion picture on the program, two banquets, three dances, a gala night, and president's reception, as well as a tea, fashion show, lunch, bridge, and a special tour of San Francisco for the ladies.

The largest display of water works and sewage treatment equipment ever exhibited appeared on the convention floor of the auditorium. There were 102 exhibitors who filled 159 booths and the auditorium floor was not large enough to accommodate them all. Several exhibits had to be set up in the corridors on both sides.

It is probably too early to fully appraise the success of a concurrent convention of this type. Without doubt from the standpoint of both the attendees and the manufacturers, the exhibits were far larger and better than could be hoped for at separate meetings. Naturally certain difficulties arose, many of which might be circumvented in any attempt of a similar meeting in the future. Comments ranged all the way from 'never again' to 'should be held every year.' Other suggestions ranged from 'holding such meetings once in five years' to 'holding the two meetings contiguously', as is now being done in some of the state organizations. Despite some criticism of detail and difficulties, no one can say that the meeting was not a success, and reflective judgment on this meeting will determine whether or not meetings of this type are feasible, practicable, and advantageous, and, if so, how frequently such meetings should be held.

The Silver Anniversary Meeting in New York in 1952 afforded opportunity for a comprehensive inventory of the progress of the Federation during its first quarter-century. The event was all the more memorable for the presence and participation of many of the founding fathers. The Silver Anniversary Luncheon speeches by Charles A. Emerson and Abel Wolman were outstanding features. Earlier chapters of this history quote liberally from Emerson's paper.

Fifteen years later the Federation again held another record-breaking meeting, its fortieth, in New York City, where the registration was 4,318. Outstanding events at that meeting were addresses by New York Governor Nelson A. Rockefeller and New York City Mayor John Lindsay, who showed a model of the new North River plant to be built to relieve pollution from upper Manhattan (The plant is still years from completion in 1977).

After 1967, except for the Dallas meeting in 1969 and the Cleveland Conference in 1973, conference registration surpassed that of the previous year, reaching a maximum of 8,716 in 1976 at Minneapolis.

Conference Technical Program

The Annual Meetings could not have been so successful had it not been for the excellence of the technical sessions. Credit for this was due the Publications Committee which had the responsibility for developing the conference program and was chaired by F. Wellington Gilcrease from its creation in 1941 until 1955.

Beginning in 1956-57, records were kept of the time allotted to various categories of program content. The summary in Table VI-2 clearly shows the growth of the Federation and its interests as well as the growth in information available for dissemination.

Two interesting sidelights on program development concern research and industrial wastes. In midsummer 1955, George Symons called Rolf Eliassen, Program Chairman, and suggested that a Research Symposium be added to the program for that fall in Atlantic City. Dr. Eliassen agreed, with two conditions: that the session be held on Thursday afternoon, after the convention had officially ended, and that George Symons choose the participants and chair the session. Thirteen researchers were solicited for papers, six responded, four were accepted. The forum was held as stipulated after the official program had closed; 120 persons attended the session. This became a successful feature of all later meetings.

The second sidelight concerns the development of industrial waste sessions. Ken Watson (Federation President, 1957-58) chairman of the Industrial Waste Committee, was made an active member of the Publications and Program Committee in 1957-58. Ken proposed more industrial waste papers and a

TABLE VI-2. Number of Hours of Daytime Programming

Topic	1962	1961	1960	1959	1958	1957	1956	1952
Total Hours	52.75	55.5	51	42	39	32	28	23.75
Business & Luncheons	3.5	4.5*	4	4*	3.5	2	2	2
Inspection Trips	3	6	9**	6	5.5	8	4	3
Sewage Treat. & Stream Pollution	5.25	9.75	3	7	5	3.75	3	0.75
Management	6.0	2.75	3	3	—	—	—	1
Design, Operation & Maintenance	13.75	12.00	15	9	9	6.75	9.75	8.75
Research	9.25	6.0	3	3.5	6	3.75	3	2.25
Industrial Wastes	10.5	11.5	12	8.5	9	6.75	5.25	6.00
Sewage Analysis	—	1.0	2	1	1	1	1	—
Movies	2.0	2.0	—	—	—	—	—	—

*Does not include WSWMA luncheon.
**Three concurrent trips.

luncheon meeting of those interested in industrial waste. The majority of the committee was not enthusiastic about giving more attention to industrial wastes, but Ken's persistence paid off and papers on industrial waste became a regular part of conference planning thereafter.

The following addendum to the 1961 Publications and Program Committee Report by outgoing Chairman Symons gives a good insight into the development of the annual conference programs during the years 1956-57 through 1960-61:

> During the past five years, the number of hours of daytime programming almost doubled from 28 to 55. This increase was necessitated by the growth of the organization, the increase in number of registrants, the expanding interests of Federation members and the increase in the number of good papers available. The increase in programming hours was accomplished by introducing concurrent sessions and extending the meeting through Thursday afternoon.
>
> Among the featured sessions introduced in the past five years were:
>
> > Laboratory Scientists Breakfast;
> > Forum on Maintenance (Plant and Sewer);
> > Industry Day;
> > Industrial Waste sessions in addition to Forum;
> > Industrial Waste Luncheon;
> > Movies for session openers; and
> > More time for luncheons
>
> Time was also provided for the Water & Sewage Works Manufacturers luncheon and the committee continued the Research

Symposium which this chairman introduced as a committee member in 1955.

Among several new ideas adopted for committee operation were the institution of a planning session held Wednesday morning at each annual Federation meeting. The mid-winter committee meeting in New York City was also continued and time-table scheduling adopted.

As a guide to programming, an analysis of meeting registration and session attendance was instituted, as was the practice of asking past presidents to serve as presiding officers at technical sessions.

We have endeavored to make our annual report informative and not just a mere statement that the committee had provided a program for the annual meeting. One effort which we believe quite worthwhile was the publication in the Journal of an explanation of how a Federation meeting is programmed.

As my term as Chairman of this Committee comes to a close, I want to thank the Presidents under whom I have served and the many Board Members who have given the committee guidance and cooperation.

The chairman and the Federation are indebted to the individuals who served on the Committee for the past five years. During that period the committee membership grew from 15 to 21, plus the ex-officio President and Vice President each year. All were working members. There were 38 persons who served on the committee during this period; six for five years, one for 4 years, four for three years, nine for two years, and 18 for one year.

The vice chairman of the Committee, Paul D. Haney has served a full five year term and now will be chairman of the Program Committee."*

Who Attended WPCF Meetings?

For many years, the equipment manufacturers association (WWEMA) retained a consultant to analyze and categorize the attendance data for each annual WPCF conference. The data for several years were made available to the Federation and are of historical interest. Table VI-3 shows the registration data breakdown by categories and Table VI-4 shows the breakdown of the registration data by percentage for the years 1958-68 inclusive. Registration data for all of the conferences appears in the Appendix.

---

*The committee was actually Publications and Program.

TABLE VI-3. Registration Breakdown by Year
(Number of Registrants)

Category	1968 Chicago	1967 New York	1966 Kansas City	1965 Atlantic City	1964 Bal Harbour	1963 Seattle	1962 Toronto	1961 Milwaukee	1960 Phila.	1959 Dallas	1958 Detroit	1952 New York
1. Manufacturers' representatives	2165	1559	1315	1295	585	528	695	720	505	349	425	258
2. Municipal employees	487	548	530	387	325	372	351	305	290	231	273	112
3. Consulting engineers	568	692	436	371	225	200	291	230	240	170	164	130
4. Industrial waste plants	198	190	195	166	105	109	93	167	138	68	72	72
5. Health Dept. engineers	206	223	251	197	145	172	184	142	138	105	94	
6. University professors and students	253	261	197	193	110	116	70	100	51	79	68	103
7. Mayors, commissioners and attorneys	108	14	49	69	35	50	61	26	10	11	17	
8. Foreign	72	44	27	43	30	27	12	12	23	22	8	
9. Miscellaneous	104	184	40	56	30	27	12	12				
Total Men	4161	3715	3040	2878	1590	1588	1774	1728	1395	1035	1111	751
Ladies	645	603	443	583	464	363	503	328	325	237	238	
Total Registration	4806	4318	3483	3461	2054	1951	2277	2056	1720	1272	1349	1152

Key

1 — Manufacturers' employees and representatives
2 — Managers, superintendents, engineers, chemists, operators (municipal)
3 — Consulting engineers
4 — Managers, superintendents, engineers, chemists, operators (industrial waste plants)
5 — Health department engineers, chemists, etc. (federal, state, local)
6 — University professors and students
7 — Mayors, commissioners, trustees, attorneys, councilmen
8 — Foreign—outside continental U.S. and Canada
9 — WPCF & WWEMA staff; retired; newspapermen

TABLE VI-4. Registration Breakdown by Years (Percentage)

	1968 Chicago	1967 New York	1966 Kansas City	1965 Atlantic City	1964 Bal Harbour	1963 Seattle	1962 Toronto	1961 Milwaukee	1960 Phila.	1959 Dallas	1958 Detroit
1. Manufacturers' representatives	45.0%	42.2%	37.6%	37.4%	28.4%	27.0%	30.4%	34.9%	29.4%	27.5%	31.5%
2. Municipal	10.1	15.7	15.2	11.2	15.8	19.1	15.4	14.8	16.8	18.3	20.2
3. Consulting engineers	11.8	18.6	12.6	10.8	10.9	10.2	12.8	11.2	14.0	13.4	12.2
4. Industrial waste plants	4.1	5.1	5.6	4.8	5.1	5.6	4.1	8.1	8.0	5.3	4.6
5. Health Dept. engineers	4.2	6.0	7.2	5.7	7.1	8.85	8.1	6.9	8.0	8.2	7.0
6. University	5.3	6.1	5.6	5.6	5.3	6.0	3.1	4.9	2.9	6.2	5.0
7. Mayors, commissioners, etc.	2.2	0.3	1.4	2.0	1.7	2.55	2.7	1.25	0.6	0.8	1.2
8. Foreign	1.5	1.1	0.8	1.2	1.4	1.4	0.55	0.6	1.4	1.7	0.6
9. Miscellaneous	2.2	4.9	1.2	1.6	1.5	0.7	0.75	1.25			
10. Ladies	13.4	13.9	12.8	22.0	22.8	18.6	22.1	15.9	19.0	18.5	17.6

Comment

a. All registrants not listed in the published lists were distributed to the various categories according to the distribution of Tuesday registrants.

b. The distribution data for 1966 are self-explanatory; the drop in ladies registration from 1965 was offset by the increase in one-day and student registrations, the latter (64) constituting about ⅓ of the total university category.

c. Kansas City and No. Kansas City, Mo., and Kansas City, Kan. accounted for 300 men registrants or 8.7 percent of the total (men and ladies) registration. These registrants were distributed as follows:

Manufacturers	89
Managers, superintendents, etc.	71
Consultants	61
Ind. waste engineers	6
Govt. engineers	22
University professors	2
Mayors, etc.	5
Miscellaneous	7
Water plant operators	37

Reconvened Conferences

Not every WPCF conference was limited to three, four, or five days within a single week. Several were extended by reconvened sessions in other locations. There were eight such meetings, held outside the coterminous states, the first in 1963 and the latest in 1975. The information on these reconvened conferences is shown in Table VI-5.

TABLE VI-5. Data on Reconvened Conferences

Conference	Date	Reconvened Conference	Date	Number Attending
Seattle	10/6-10/63	Honolulu, Ha.	10/13-16/63	500
Bal Harbour, Fl.	9/27-10/1/64	San Juan, P.R.	10/4-7/64	98
New York	10/8-13/67	San Juan, P.R.	10/15-18/67	247
Dallas	10/5-10/69	Monterrey, Mex.	10/12-14/69	400
San Francisco	10/3-8/71	Honolulu, Ha.	10/10-13/71	325
Atlanta	10/8-13/72	San Juan, P.R.	10/15-18/72	400
Cleveland	9/30-10/5/73	Toronto, Can.	10/6-9/73	150
Miami	10/5-10/75	San Juan, P.R.	10/12-15/75	375

Education Program

In New York October 11, 1941, the Board of Control appointed an Operator Qualifications Committee and charged it with establishing minimum qualifications for operators of various classes of treatment works and to promote acceptance of these qualifications on a voluntary basis.

In general the aim of this activity was to maintain desirable standards for sewage plant operators. The following year the committee report, by Chairman H.G. Baity, to the Board of Control contained the following objectives: "(a) to develop a code of sound and responsible qualifications for operators of various types and sizes of sewage treatment plants and to suggest the acceptance of such uniform qualifications on a national basis; (b) to study the various systems of regulations now in effect in various parts of the country, such as civil service health department qualifications, voluntary licensing plan, etc., and to suggest means of achieving substantial compliance with uniform standards in all states."

Various subsequent reports of the committee provided information to state operator training and certification programs and were aimed at promoting these programs throughout the states. These early surveys and reports provided a base for subsequent committees to continue their work of determining current practice on operator training and certification, and promoting improved certification programs.

In 1955, a resolution was introduced by the committee to the Board urging all Member Associations to emphasize the importance of securing and retaining competent personnel in treatment plants. In 1959, the title of the committee was changed to the Personnel Advancement Committee.

The 1960 Personnel Advancement Committee report in the *Journal* included a voluntary certification plan and listed programs for training and certification by states.

In 1963, there was also a joint certification committee with representatives of AWWA, WPCF, the Conference of State Sanitary Engineers, and the U.S. Public Health Service working to clarify the differences between the recommended mandatory certification programs of AWWA and WPCF. The resulting model law for both water and wastewater was published in the December 1966 issue of the *Journal*.

Wastewater treatment plant training course and aids.

Beginning in 1964, the Personnel Advancement Committee started to develop training materials as well as to collect data on licensing, certification, training, and salaries. In the same year, the Personnel Advancement Committee prepared, with staff assistance, the drafts of Training Course I lecture outlines and color slides, that were published and made available for use by Member Associations and others.

Course I, consisting of visual aids and a series of lecture outlines, sold for $25 per set. The Board of Control also recommended at that time that the Federation staff involve itself more in the promotion of training. This was a period in the Federation's history when outside funds were solicited.

Federal grants were obtained in the next few years and were used to develop Course II slides and lecture outlines, Safety Promotional Materials, to sponsor an operator training workshop held in Dallas in 1969, and even as late as 1972 to aid in the development and publication of the MTM-01 Extended Aeration Training package. During this period a number of projects were completed with staff assistance and outside consultant contractors.

During the period of using funds from outside the Federation to support special projects, there was one dealing with funding from WWEMA to provide a stipend and travel expenses for treatment plant operators to attend short courses and other types of pertinent training programs available. In 1967, WWEMA offered $10,000 to AWWA and WPCF to establish scholarship funds for operator training. This figure was increased later to $10,000 per organization. A program to distribute these funds to operators throughout the U.S. was proposed and put into operation. The funds were to be used for training the maximum possible number of operators and travel fees and subsistence, and were to be distributed on a wide geographical basis.

The Personnel Advancement Committee's Report of 1969 stated: "To date 140 grants totaling $14,325 have been awarded. Forty-five of these were awarded in 1969. If the 1968 experience is any indication of what will happen in 1969, we should end this year with a 2-year total of 250 operator grantees receiving $25,000."

The program was highly publicized in *Highlights* and special mailers were sent out to municipalities and state agencies. Applications for training grants came in from all over the U.S.

Because of national interest and the need for information on manpower and training, the Federation formed a Man-Power committee in 1967 as requested by President Arthur Caster and Vice-President Berkowitz. The committee was formed in June 1967 and given the following charge: "The committee should (a) prepare the format for and commence a survey of the total manpower needs in the water pollution control field, including all disciplines and interests; (b) on the basis of personnel needs and trends, develop projections of personnel needs for the foreseeable future; and (c) submit a report of progress to the Board of Control at the 1968 Conference including therein a plan for publication and distribution of the findings and the projections together with a plan for maintaining and updating data no less often than biannually." Shortly afterward it was agreed to add salary surveys to the charge.

The Board of Control in 1969 voted to co-sponsor an operator training workshop with Clemson University to be held in Atlanta. This workshop was also to be co-sponsored by the Federal Water Pollution Control Administration and was directed at training program instructors, while the operator workshop that was held by the WPCF in conjunction with the WPCF Annual Conference in Dallas in 1969 was operator-oriented. The Clemson workshop followed in November, 1969.

At the Dallas operator workshop, a resolution was passed and directed to the Board of Control that had a distinctive impact on the educational program of the WPCF. The resolution stated: "We recommend to the Federation Board of Control, as a matter of high priority, that a national instrument be established to formulate a national plan for, and to give leadership to, the implementation of programs in the broad area of operator recruitment, training, and certification; and notification of this action be transmitted to AWWA and such other appropriate organizations and agencies as may be desirable for their action and collaboration." This resolution resulted in a study by the staff and members of the Personnel Advancement Committee into the development of what was referred to at a later date as a MANFORCE (MANpower FOR a Clean Environment) document.

In July 1970, an additional staff member was employed to act specifically as a Manager of Education and Training, his primary purpose being to guide the activities of "MANFORCE." Funding was a major problem, and one of the major objectives of the new Manager of Education and Training was to develop outside sources of funding. Considerable time was spent in this area.

The Personnel Advancement Committee in 1970 discussed the need for national action on operator certification. As a result, selected individuals representing state certification programs were invited to discuss what type of national organization might be desirable and possible. This meeting led to the organization of *ad hoc* committees to form ABC (Association of Boards of Certification for Operating Personnel in Water and Wastewater Utilities). The Personnel Advancement Committee chairman and a member were selected as the representatives from the Federation to the ABC Executive Committee. They were Heinz Russelman and Terry Regan.

The Personnel Advancement Committee cooperated in a survey of state certification programs. The results of this survey were published in the journals of both the Water Pollution Control Federation and the American Water Works Association.

One of the results of the MANFORCE program was MTM-01 (Manforce Training Module 01) which is an Extended Aeration Training Module. This training package, designed for operators of extended aeration treatment plants of 50,000 gallons per day or less was a major step in the modern approach to training. The material was designed in such a manner that it may be used for self-study, in a classroom setting, or as reference material for instructors' purposes. The material went far beyond what Course I and Course II did in providing only slides and lecture outlines.

About this time the Federation Board decided that the use of federal grants for its programs was to be discouraged, and after 1972, no further federal grants were solicited or accepted.

A task force was established within the Personnel Advancement Committee to study MANFORCE and review whether it should be redirected toward working with Member Associations, for example, in assisting with developing workshops and other training endeavors for their members.

At the Federation Conference in Cleveland in 1973, Ed Braatelien, Jr. was appointed chairman of the Personnel Advancement Committee and a new charge was developed for the committee: "The committee is charged with the responsibility to develop and increase technical competence of personnel who are engaged in wastewater management programs, in order to maintain and enhance water quality and to achieve justifiable recognition of persons working in the field through the support and coordination of activities of individual members and Member Associations of the Federation."

The first priority under the new charge was a training survey to determine what involvement each Member Association had in education and personnel advancement through its members. This survey was published in the February 1976 *Deeds & Data*. The Personnel Advancement Committee completed a certification examination studybook started under the MANFORCE program. The questions had been collected from state agencies and other agencies that provided certification examinations for wastewater treatment plant operators throughout the U.S. and Canada. Originally, there were about 3,600 questions and after duplications and ambiguous questions were eliminated the Studybook eventually was printed with over 830 questions that represented the various certification examinations in the wastewater field.

Major emphasis of the new Personnel Advancement Committee was placed on the annual pre-conference workshop. The workshops were oriented toward advancing treatment plant operators, managers, and engineers involved in operational problems.

Starting with the Atlanta Conference in 1972, the pre-conference workshops were not only self-supporting but also very successful technical meetings covering operational problems.

The Cleveland Pre-conference Workshop had as its main theme the Occupational Safety and Health Act (OSHA). There were 86 participants in Cleveland and the materials generated for this meeting stimulated OSHA workshops in some Mem-

Initiated at the 1972 Atlanta Conference, the Pre-Conference Workshops have proved to be self-supporting as well as successful meetings that cover operational problems.

ber Associations (Rocky Mountain, Virginia, Chesapeake, and Michigan).

The Denver Pre-conference Workshop in 1974 covered "The Ways and Means of Personnel Advancement." There were 102 participants, and some Member Associations went on to hold their own workshops on operator training and advancement.

The Miami Pre-conference Workshop in 1975 covered "Emergency Action Planning" and had 136 participants. Both Member Associations and the EPA sponsored a number of workshops and seminars after the Miami meeting.

The Minneapolis Pre-conference Workshop in 1976 was on "Operator Training Methods and Media." It had 192 participants and was highlighted with the introduction of the WPCF new "Basic Course." The Personnel Advancement Committee was committed to continue the pre-conference workshop covering topics that will meet the future needs of operational personnel in the water pollution control field.

Member Associations sponsored many other workshops that were not a direct result of pre-conference workshops. Examples were safety seminars for supervisors, management and labor relations workshops, PL 92-500 Sections 201 and 208, planning seminars, operation and maintenance manual preparation, hazardous spills, and the coastal impact seminar.

The Federation cooperated in the production of an audio-visual public information package originally developed by the Los Angeles Section of the California Water Pollution Control Association. The program was an introduction to a wastewater treatment plant tour for junior and senior high school students. Titled, "Have You Ever Wondered?", it was introduced at the 49th Annual Conference in Minneapolis in October 1976, and was distributed to all Member Association Secretaries for the use and benefit of their members.

At the Miami meeting in 1975, the Personnel Advancement Committee reached an agreement with Environment Canada on the joint development and production of the WPCF "Basic Course." This material had originally been developed in Canada for training entry level treatment plant operators in the Maritime Provinces.

The Personnel Advancement Committee established a task force to review the material and the technical content as well as the method of presentation. The material was produced and released to the public at the annual meeting in Minneapolis, October 1976. Thereafter, the Personnel Advancement Committee and Educational staff continued to work with Environment Canada on a Level II, intermediate course which will be a continuation of the systemized program approach to training on which the basic course was developed.

In 1976, the Chairman of the Personnel Advancement Committee was appointed to a steering committee formed with AWWA/ABC and WPCF on the development of a model training and certification program for the U.S. and Canada. This committee published a report—referred to as the ABC "Brown Book,"—and if all the recommendations are accepted by the water pollution control field, future training and educational programs will be coordinated at the state level and the transfer of information will be meaningful on a national level.

The latest involvement of the Personnel Advancement Committee and the Education staff was a 1976 survey of wages and fringe benefits for wastewater facility personnel covering both collection systems and plant personnel. This document was scheduled for publication in 1977.

Technical Services

From 1928 on the Federation had been a source of technical information on the control of wastewaters and pollution, and providing that information continued to be the primary objective of its *Journal* and various publications. After 1941, the Federation maintained a staff of professionals to provide a reservoir of technical information for its membership and others making specific inquiry. Until 1965, however, the Federation did not have a staff commitment exclusively for conducting technical programs and responding to technical inquiries. Before that time, members of the editorial staff assisted the Executive Secretary in providing technical response and supporting committee efforts.

During 1963 and 1964, discussions on expanding WPCF staff to meet the increasing need for technical services, such as operator training and certification and safety program promotion, led to the 1964 Board approval of a technical services manager and committee liaison positions as recommended by a special committee on staff study. Only one position was budgeted for 1965, however, and a professional staff member was recruited late in that year to start a program combining these two functions.

Early in the program development, two grants from the Public Health Service were pursued and obtained to develop materials for operator training and to promote safety and operator certification.

At the same time, the technical services program provided staff response to technical as well as lay inquiries about water pollution control matters. A collection of water pollution control leaflets from the federal government and from private organizations was assembled for use in responding to various inquiries. Also, various pamphlets were developed for complimentary distribution for teach-ins, environmental meetings, and debates, and for informing school pupils who made direct inquiries. A pamphlet on water pollution control careers was produced for the use of school career counselors and the general public.

The first use of the careers pamphlet was at a special meeting of students with representatives of AWWA and WPCF at the AWWA conference in Cleveland in 1968. The document became very popular, in part because of its use on many source lists furnished to school teachers by various agencies. For a while it was used by the federal civil service offices. Later the document was revised to include additional information, and in 1976 it was reviewed by the Federation Human Resources Committee in order to include many additional disciplines which began to enter the field in the 1970's.

Wastewater Treatment Plant Operator Training Course Two was finished and released in 1967, and quickly became popular in the field. A contract study on operator certification was conducted during 1967 in order to obtain an outside view of the programs and their benefits. A report of the study was published in the *Journal* for January 1967. This report and the annual surveys of operator training and certification indicated increasing momentum toward mandatory certification. This movement was assisted by the various Federation publications on the subject and the joint publication of their recommendation for requiring certification of water and wastewater works operators, published in December 1966 by the Federation and the AWWA in their respective journals.

In the early 1970s, the federal water pollution control pro-

The Federation's career pamphlet that was used extensively by career counselors throughout the United States and Canada.

gram increased its interest in the state certification programs and provided assistance in developing model legislation. On request of EPA, the Federation assisted in presenting to the Council of State Governments a proposed model act for their consideration and endorsement. This effort resulted in the council including a model act in its 1973 recommended state legislation that was basically the same as the 1966 model but with some additions and adjustments. A report on this and the model act was published in the October 1972 *Journal*.

On the urging of a few state and other officials having responsibility and concern for certification programs in the states and the role of these programs in the national water pollution control program, the Federation in January 1971 hosted a meeting of selected state officials to discuss the possibility of organizing certification program agencies on a national basis. Subsequently, this group expanded, and with the help of the Federation and the AWWA, organized the Association of Boards of Certification for operating personnel in Water and Wastewater Utilities (ABC) in Chicago in June 1972. In recognition of the close relationship of this organization with the Federation and AWWA, the ABC Constitution established positions on its Executive Committee for two representatives from WPCF and two from AWWA. To assist ABC in getting started, the Federation furnished a secretary for staff services for several years and financial contributions were still being made in 1976-77.

In 1971, EPA asked the Federation to consider contracting a special state certification program survey and developing a model program. A contract was signed, and through a consultant a study was made in 1972. The results of this study were published in the May 1972 *Journal*.

Several years of deliberations by the Safety Committee in the early 1960's resulted in reports to the Board of Control in 1964 and 1965 calling for a full-time staff member to conduct a safety promotion program. With the addition of the staff member as Manager of Technical Services and Committee Liaison in 1965, and the approval of a federal grant in 1967, an organized program to promote safety was begun. With the formation of a special task force in the Safety Committee and the assistance of the National Safety Council's Association's Director, a special survey of wastewater works was conducted to learn of their safety programs, their needs, and the extent of injury in their operation.

Based on the information produced by this survey, and with the services of a professional consultant on safety, promotional materials were produced. These were released in packet form with supplementary slides at the annual conference in Dallas in 1969. During 1970, two workshops were conducted to counsel representatives of Member Associations on the use of the safety packet to promote safety programs in their respective areas. The first 4,000 safety packets were made available through the Associations on a complimentary basis for this purpose.

Promotion through the Member Associations was effected by *Journal* and *Highlights* articles, special bulletins, and the planning and furnishing of safety program materials. A special one-day safety seminar for supervisors was developed for the use of Member Associations. It included lecture material, motion picture film, flannel boards with signs, a demonstration model, and guidance on organizing and conducting the seminar. Assistance was also offered in locating or furnishing speakers and helping in advertising and promotion. To introduce this package of material, the seminar was conducted at the Boston Conference in 1970.

At the Boston Conference, the Federation received the National Safety Council's Association Safety Award in recognition of distinguished service in promoting the efficiency and effectiveness of safety programs. At the same conference the Federation made its first presentation of its new Member Association Safety Award to the Indiana Water Pollution Control Association and gave honorable mention to the Chesapeake Water Pollution Control Association for their safety efforts. In a special commemorative edition of the Boston Herald Traveller, the Water Pollution Control Federation Conference was featured. A front-page article in this edition reports on the Federation being cited by the National Safety Council and of the Federation presenting its Association Safety Award to the Indiana Water Pollution Control Association.

The organization of safety programs by Member Associations and the increasing response to the annual safety surveys indi-

The Federation's
Safety
Promotional
Packet
and
slides.

Gordon L. Burt, left, Chairman of the WPCF Safety Committee, accepts the 1970 Association Safety Award from the National Safety Council presented by Cole A. Allen of the NSC Board of Directors. The NSC award recognizes outstanding work by WPCF in promoting safety programs among its Member Associations and membership and in the development of instructional material for these programs. The award was presented at the Federation Luncheon on October 6 in Boston.

A checklist of useful sewer maintenance equipment includes foul weather suits and boots; rubber and canvas gloves; safety harness and vest; traffic cones; flashing lights; canister and self-contained masks; rope; blowers; and explosimeter.

cated that the Federation's promotional efforts were beginning to succeed. Records of the annual safety surveys in the early 1970's show that injury rates were beginning to decline. However, beginning in 1973, this trend reversed and by 1974 there was an alarming increase in injury rate among collection system workers also. Based on these warnings, new and increased efforts, including the reorganization of the Safety Committee in 1976, were planned.

Manuals of practice, as originally conceived and currently produced, have always been the products of committees of experts in their respective fields who draft, review, and rewrite technical documents for approval by the Technical Practice Committee. The manuals present acceptable practice in particular segments of the water pollution control field. Not all MOP's however, were actually written by committees; for example, the "Fertilizer Manual," the forerunner of what is currently MOP 2, was written by Langdon Pearse; and the "Air Diffusion Manual" was contributed almost solely by Norval E. Anderson. The

"Sewer Ordinance Manual" was prepared in draft form by Secretary Wisely for review and clearance by the Committee.

For the most part, the usual practice for preparation of the individual manuals began with the selection by the Technical Practice Committee of an appropriate subcommittee chairman to oversee the preparation (or revision) of a manual. Subsequent procedures included the subcommittee chairman's solicitation of subcommittee members suitably qualified to assist in the task by drafting appropriate sections of the manual, followed by review of the completed manual by the full Technical Practice Committee, and outside reviewers, if appropriate. Throughout the drafting and review process, staff services were furnished as required. This procedure was generally followed throughout the history of the Manuals of Practice development and, aside from occasional delays, worked well.

The rapidly increasing efforts to improve water pollution control in the late 1960's resulted in more technical and lay literature being published. Increases in federal program grants brought about increasing demand for regulations and guidance on compliance. The Environmental Protection Agency's publication of its guidelines on design and operation and maintenance at the Federation's annual conference in Boston 1970 brought about a review and discussion of the roles of the Federation and the federal government for disseminating information on technical practice. It became obvious to the Federation that its past practice in producing Manuals of Practice would have to be revised in order to meet the increasing demand for the latest information on current practices.

The Federation began seeking volunteer authors who were expert in their respective fields. The Technical Practice Committee, made up of these experts, was composed of as many as 15 to 20 subcommittees, each responsible for a particular document's preparation or revision. The membership of the committee and its subcommittees varied from year to year, depending upon the state of activity, and frequently exceeded 100 people. Usually staff assistance was limited to final editorial activity and the usual management of printing. Experience, however, began to show that the pressures placed upon authors prevented them from drafting manuals as fast as the increasing demand for the material.

The Technical Practice Committee gave serious consideration to its concept of MOP preparation in 1972. This study concluded that the technique of using voluntary input by experts in various segments of the field was still the best way of producing documents on current acceptable practice. In order to continue use of this technique while speeding up the production of the documents, the Committee recommended an increase in Federation staff to support the Committee. By action of the Board, the 1973 budget was increased but full implementation of the procedure after staff recruitment did not take effect until 1975-76. Also, the Committee concluded that although the potential for duplication was great, the federal agencies and the Federation, through close coordination, would be able to meet their respective objectives by continuing to produce publications which were being expected of them.

The addition of an equivalent of one professional staff member and clerical support in 1973 enabled the staff to participate in the work of the authors, review drafts, analyze review comments with the chairman, and rewrite manuals for subsequent review. With gradual adjustment of the role of the staff and committee and subcommittee members, the new procedure resulted in the completion of two revised manuals in 1975, and four revised manuals and one new manual in 1976. Total new and revised pages issued was 874. An eighth document with over 550 pages was scheduled for publication in 1977.

Even with what was obviously a successful revision procedure during the mid-1970's, the Technical Practice Committee continued to study further modifications which would enable it to meet the continuing increased demand for technical publication. A special subcommittee, under Federation Past President Joe Lagnese, worked during 1976 and in its report at the annual conference in Minneapolis, recommended a restructuring of the committee and a revision of the breakdown of subject matter for manuals. This report, which was approved by the committee, called for more manuals with each covering a smaller segment of the field.

Inasmuch as the subcommittee was unable to delineate all fields and details of procedure, it recommended the establishment of three technical subcommittees, one each in the following categories: facilities development (design), operation and maintenance, and management. These subcommittees would be responsible for planning the preparation of manuals and delineating the respective subjects to be covered. In its recommendation, the subcommittee also included maintenance of liaison with government agencies and other professional associations and a strong support by the staff to assist the task forces which would work under the direction of the three technical subcommittees.

Concurrent with the reorganization of the Technical Practice Committee, staff support was provided by Technical Services staff consisting of three professional engineers, a secretary and an information clerk. Thus, with a revised and modernized approach to manual preparation and expanded technical staff, the Federation, at the end of its first half century, could look forward to greatly accelerated preparation of manuals of practice and other technical publications.

Government Affairs

The history of the Federation's government affairs program dates to 1947 when the Legislative Committee was established to study legislation pertinent to the interests of the Federation. Thereafter, this important activity evolved gradually but steadily. In large measure, it mirrored the increasing involvement of the federal government in the water pollution control field. In effect, one can only address the Federation's increasing interest in and commitment to government affairs in terms of the history of the federal water pollution control program, including the passage over a Presidential veto of PL 92-500, the Federal Water Pollution Control Act Amendments of 1972.

While the history of the federal response to the dangers of water pollution dates back to the nineteenth century, the first legislative proposals to create a separate federal program were introduced between 1936 and 1948. Similarly, the Federation's response to this initial legislation dates from the same time. In this regard, Ralph E. Fuhrman, first Chairman of the Legislative Analysis Committee and later Federation President (1951-52) and Executive Secretary, presented for consideration and adoption by the 1947 Board of Control the following statement of policy:

> The Board of Control of the Federation of Sewage Works Associations hereby records the policy of the Federation on federal water pollution control.
>
> The Federation favors federal legislation:
>
> (1) That will use to the greatest possible extent existing authorities and facilities for the control of water pollution;
>
> (2) That designates the U.S. Public Health Service as the federal Administrative Agency;
>
> (3) That will provide, in the classification of waters, for cooperation and agreement of and between the U.S. Public Health Service and the state pollution control authority, and where no such state authority exists, the Surgeon General of the U.S. Public Health Service shall integrate the stream classification of such states with the classification of the streams in adjoining or affected states; and
>
> (4) That provides for an Advisory Board with adequate representation of State Water Pollution Control Authorities.

In the discussion that followed the seconding of the motion, most were favorable to the proposed policy and, upon the call for the question, the statement of policy was adopted as presented.

Federal Program Begins

The Water Pollution Control Act of 1948 marked the beginning of the federal water pollution control program. That legislation encompassed the general policies that the Board of Control had adopted in October 1946. Generally, the legislation assigned to the federal government a very secondary position in water quality matters, the principal federal responsibility being to bolster state and local water pollution control programs with research support and technical and financial assistance. More specifically, the legislation:

• Directed the Surgeon General of the U.S. to assist in and encourage state studies and plans, interstate compacts, and the creation of uniform state laws to control pollution;

• Supported research with a special appropriation for the erection of a research facility;

• Authorized the Department of Justice, after notice and hearing and only with the consent of the state, to bring suits to require an individual or firm to abate pollution;

• Established the Federal Water Pollution Control Advisory

Board for the purpose of reviewing policies of the Public Health Service and making recommendations for improving procedures;

- Provided authorization for funding of $22.5 million per year for fiscal years 1949-1953 for low interest (2 percent) loans for the construction of sewage and waste treatment works; $1 million a year for fiscal years 1949-1953 for grants to states for pollution studies; and $800,000 a year for fiscal years 1949-1953 for grants to aid in drafting construction plans for water pollution control projects.

Public Law 82-579, which extended the provisions of the 1948 Act for an additional three years through fiscal year 1956, provided persons involved in the water pollution control field with additional time to become accustomed to the new federal-state-local partnership. Although the federal program was funded only meagerly during its inception, the Federation maintained its interest in and continued to support the new program. The 1949 Board of Control adopted a resolution opposing a bill pending in the 81st Congress that, from the Federation's viewpoint, would have seriously altered the state-federal relationship set forth in PL 80-845 by giving the federal government authority to bring suits in the federal courts against municipalities and industries without the consent or approval of state pollution control authorities.

During the early 1950's, the Federation Legislative Analysis Committee, under the chairmanship of David B. Lee and R.S. Shaw, continued its efforts placing primary emphasis on the review of the model state pollution control law as developed by the Federal Division of Water Pollution Control as well as on the analysis of state and federal legislation of interest to the Federation. Of particular interest at this time was the adoption by the 1950 Board of Control of a resolution emphasizing the need for Congress, in the enactment of legislation for the defense of the nation, to be fully apprised of the basic importance of pollution control measures. The resolution urged the continued construction and operation of pollution control facilities during the period of emergency by allocating necessary materials and equipment and appointing a qualified representative in the sewage and industrial wastes field to advise the National Securities Resources Board, the National Production Authority, and other responsible agencies with respect to such allocations.

In 1956 the Congress, through the enactment of PL 84-660, placed the federal water pollution control program on firm footing. While this legislation reaffirmed the principle that the federal role should supplement state pollution control activity, sizable funding increases and stiffer federal enforcement powers indicated that the federal government's involvement in the water pollution control field was being significantly expanded. More specifically, the Water Pollution Control Act Amendments of 1956:

- greatly strengthened the research and training aspects of the federal program;

- provided $3 million a year in grants for fiscal years 1957-1961 to assist in the preparation of state and interstate pollution control plans;

- replaced the construction loan program of the 1948 Act with a $50 million a year grant program to assist local communities to build sewage treatment plants; and

- strengthened federal enforcement procedures through the removal of the state consent requirement.

After passage of this legislation, the Legislative Committee intensified its efforts, and federal water pollution legislation became the Committee's principal interest. In 1956, Chairman V.W. Bacon urged meetings to assist various state and other officials with the detailed operation of the new law. In this regard, the new construction grant program was watched very closely inasmuch as it was included in the 1956 Act over the objections of a majority of engineers in state regulatory agencies.

Despite this initial opposition, the Federation soon supported the new law. At the 1958 annual meeting of the Federation in Detroit, Director Stiemke presented a resolution endorsing PL 84-660 and calling for increased federal funding under the new construction grant program. In moving adoption of the resolution, Director Stiemke indicated that for the Federation not to endorse the resolution would seem to refute the obvious need for increased pollution abatement activity and be contrary to the avowed purpose of the Federation. Some objections were raised by Board members, but the following resolution on PL 84-660 was passed with only two dissenting votes:

Whereas,
the increasing demands on the beneficial over-all utilization of the nation's limited water supply emphasizes the urgent need for concentrated attention to more rigid water quality regulations; and

Whereas,
water resource and river basin development requirements are on the increase throughout the nation; and

Whereas,
such projects, when completed, might materially affect the water quality conditions of the entire basin; and

Whereas,
water pollution control activities are directly related and include such water quality considerations; and

Whereas,
the enactment by the 84th Congress of the Federal Water Pollution Control Act has been successful in stimulating this local-state-federal action program, resulting in the achievement of the nation's all time high rate of construction of sewage treatment plants; and

Whereas,
in spite of this all-time high rate of construction, it still falls far short of the $575 million annual rate needed; and

Whereas,
the progress already noted in sewage treatment plants construction would be greatly accelerated by proposed amendments by Section 6 of the Water Pollution Control Act increasing the total appropriations authorization to $1 billion, increasing maximum grant provisions to projects to $500 thousand, authorizing municipalities to band together in construction of joint projects and permitting reallocation of unobligated funds; and

Whereas,
national, state and regional groups interested in the field of water pollution control have evidenced approval of such amendments in public hearings; therefore, be it

RESOLVED,
that the Federation of Sewage and Industrial Wastes Associations in conference assembled on October 9, 1958 at Detroit, Michigan fully endorse the principles of the Federal Water Pollution Control Act, Public Law 660, 84th Congress, and the cooperative local-state-federal program it supports; and be it further

RESOLVED,
that this Federation also endorses amending Section 6 of said

Act to provide additional stimulus and more effective means of reducing the backlog of pollution abatement needs; and be it further

RESOLVED,
that a copy of this resolution be forwarded to the President of the United States, the President of the U.S. Senate, the Speaker of the U.S. House of Representatives, the Secretary of Health, Education, and Welfare, the Chairmen of the House and Senate Public Works Committees, the Chairmen of the House and Senate Appropriation Committees, the Surgeon General of the Public Health Service, the governors and water pollution control agencies of the 49 states and the territories of Hawaii, Puerto Rico, and the Virgin Islands, and the members of the President's Water Pollution Control Advisory Board.

Pollution Control Act of 1961

The period 1958-1963 was particularly active both in terms of Congressional and Federation activity. With the provisions of the 1956 Act due to expire, the Congress passed PL 87-88, the Federal Water Pollution Control Act of 1961. This law made major changes in the federal program and stimulated the Federation through its General Policy Committee and Legislative Analysis Committee to adopt major policy statements and recommendations with regard to federal water pollution control activities. As enacted, PL 87-88:

- transferred administrative responsibility for the program from the Surgeon General to the Secretary of the Department of Health, Education and Welfare;
- provided for inclusion in federally sponsored reservoirs of capacity for water quality control;
- intensified federal research activities;
- raised the authorization for program grants from $3 to $5 million annually;
- raised the dollar ceiling on construction grants to $600,000 and provided for an allocation of cost in multi-municipal projects among participants with a ceiling of $2.4 million for any one project;
- raised the authorization for construction grants on a sliding scale for a period of six years to a maximum of $100 million per year;

- expanded the abatement program to include navigable or interstate waters in or adjacent to any state or states; and

- altered federal enforcement abatement procedures to provide different rules for interstate and intrastate situations.

The Federation monitored the development of this legislation and also devoted a great deal of attention to the development of a statement of policy calling attention to the need to ensure the continuation of a strong water pollution control effort. The 1960 Federation statement of policy (revised in 1962), was an important document inasmuch as it put the Federation on record as believing that:

> Pollution of the nation's watercourses, coastal waters and groundwaters is a continuing threat to the national health, comfort, safety, and economic welfare. National survival, in terms of future urban, industrial, and commercial growth and prosperity, dictates the protection of all water resources from discharges of pollution wastes and other substances, or from any acts which cause unreasonable impairment of water quality, and adversely affect their highest level of usefulness. While considerable progress has been made in pollution control by municipalities and industries, many water resources areas are being degraded, impaired, and damaged by such discharges and acts, and they will be further adversely affected by the degree and pattern of population growth, industrial processing, commercial expansion, chemical usages, and other technological advancements. (The complete Statement of Policy appears in the Appendix).

In 1963, following revision of the Federation Statement of Policy, David H. Smallhorst, Acting Chairman of the Legislative Analysis Committee, presented his Committee's report to the Board stating that the Federation, through the adoption of its Statement of Policy, assumed responsibility to the public as well as its membership for aggressively pursuing legislative matters. The report also stressed the importance of the Federation's taking a definite position with respect to proposed legislation to fulfill the obligations assumed in the statement with a recommendation to the Board to authorize the establishment of procedures for the Federation to be represented with regard to national water pollution control legislation. The Legislative Analysis Committee recommended and the Board of Control approved on Oct. 10, 1963 in Seattle, Wash., the following statement:

The Legislative Analysis Committee is charged with the responsibility of studying legislation pertinent to the interests of the Federation and serving the Board in an advisory capacity.

In carrying out this charge the Committee wishes to call attention to the fact that the WPCF in adopting its "statement of policy" assumes responsibility to the public, as well as to its membership, for pursuing in an aggressive manner the objectives so stated.

In connection with the current proposed legislation which has been reviewed during the past year and which, it is understood, is still under consideration but possibly changed in some respects, the Committee recommends that the position of the WPCF be the following:

1. With regard to the establishment of a national policy on water quality preservation, the WPCF recognizes the need for controlling the discharge of pollutional wastes into the waterways of the nation and, to this end, that decisions as to the type and degree of treatment, and control of wastes must be based on thorough consideration of all the technical and related factors involved in each portion of each drainage basin.

2. With regard to the administration of the federal water pollution control program, the position of the WPCF is that the primary objective of pollution control is the protection of the health of the public. This is reflected in the language of the Federal Water Pollution Control Act stating that pollution is measured by its effect on public health or welfare. Consequently, the WPCF believes that the Public Health Service has, by virtue of its long experience in dealing with all facets of water pollution control, including the protection of public health, demonstrated that it is best fitted to administer national water pollution control functions, and, therefore, urges that administration of water pollution control at the federal level remain with the United States Public Health Service.

3. Concerning the proposal to allocate federal funds to abate pollution caused by existing combined sewers, the WPCF takes the position that if this situation is of sufficient significance to warrant national recognition then the problem should be more thoroughly investigated as to the engineering and economic aspects that might be involved. The WPCF believes that the separation of sewer systems relates to one specific engineering solution, whereas there may be other local acceptable alternatives for the control of this pollution problem.

4. The WPCF recognizes the desirability of determining uniform water quality criteria for specific uses; however, because of

the differences in the needs of specific river basins, the WPCF recommends the establishment and use of such criteria as a cooperative effort by industry, state, local, interstate, and federal agencies for and within specific river basins.

The ad hoc committee recommends that the Board of Control instruct the President and Executive Secretary to implement the recommendations contained in this report, and to oppose any legislation or provisions thereof contrary to the Statement of Policy and/or the recommendations contained herein.

It is recommended that this report be approved and made available to the general membership at the earliest practicable date.

It became apparent within a short time that the Federation Statement of Policy, along with the recommendations of the Legislative Analysis Committee, proved extremely valuable to the Federation's involvement in legislative affairs. In 1964, in Chairman Smallhorst's absence, Past President Harry E. Schlenz observed that the Federation had accepted the responsibility for being the spokesman for the water pollution control field through the testimony of the Executive Secretary before congressional committees. Such testimony proved helpful in composing and modifying pending legislation. Schlenz, however, pointed to a committee report adopted during the conference calling for guidelines for the committee's future deliberation and action and defining the duties of the committee:

1. To obtain, review and comment on proposed legislation associated with water quality and/or water resource planning and development, including such proposed legislation involving research and education.

2. To transmit Committee reaction to the Executive Committee of the Water Pollution Control Federation.

3. To recommend, where appropriate, a policy of action of the Water Pollution Control Federation on each proposed measure."

From 1965 to 1968, the Legislative Analysis Committee, chaired by A.F. Vondrick, labored diligently to keep pace with the ever-increasing legislative activity in the Congress and made repeated efforts to improve the effectiveness of the Committee in view of the increased workload.

The Congress for its part legislated major changes to the existing legislation both in 1965 and 1966. The Water Quality Act of 1965, PL 89-234, placed primary responsibility for the administration of the program in the newly created Federal Water Pollution Control Administration in the Department of Health, Education and Welfare; raised the construction grant authorization from $100 to $150 million, and provided for a 10 percent increase in the federal share, set at 30 percent in 1956, if a project conformed with a comprehensive plan; authorized $20 million annually for research and development grants relating to methods for dealing with combined sewer pollution problems; set a timetable for the establishment of water quality standards by the states for interstate waters; and extended enforcement initiatives to abate pollution adversely affecting shellfish in interstate and navigable waters under certain circumstances.

The 1966 Act

The Clean Water Restoration Act of 1966, PL 89-753, made equally important changes in the program. This legislation increased construction grant authorizations to a total of $3.55 billion between FY 1967 and FY 1971 while removing the dollar ceiling on grants; provided for the increase of the federal share to 55 percent, if various requirements were met; authorized a new reimbursement program; raised federal participation limits as well as total authorizations for several existing research and training programs and introduced two new programs to support waste treatment research; increased authorization for state program grants; and established federal enforcement machinery with relation to international boundary waters.

The Federation responded quickly to this new legislation. In 1965, Chairman Vondrick urged the adoption of the Committee report directing the Executive Secretary of the Federation to write a letter to the new Administration offering the cooperation and support of the WPCF. The report also commended the Congress for the increased emphasis that it placed on water pollution control through the passage of the 1965 Act and urged the immediate appropriation of the additional $50 million authorization for construction grants as a means to accelerate the construction of much needed treatment projects. With regard to the function of the Committee, moreover, the report recommended changing the name of the committee to "Legislative Committee" and proposed the creation of a subcommittee to

President Johnson, seated, completes signing of Water Quality Act of 1965. Others shown, standing left to right, are Rep. Robert E. Jones (D.-Ala.), Rep. John A. Blatnik (D.-Minn.), Rep. George H. Fallon (D.-Md.), and Senator Edmund S. Muskie (D.-Me.).

study the responsibilities, channel of communications, and operational procedures in order to improve the effectiveness of the Committee.

In 1966, the recommendations developed by the subcommittee were considered and adopted by the Board. Chairman Vondrick stated to the Board of Control that the Legislative Committee was not geared to operate effectively under the circumstances of 1965 when there was an excessive number of bills considered by the Congress. He also expressed disappointment that the Executive Committee had not considered the recommendation for improving the effectiveness of the Committee. As a result, he recommended a different procedure for the next year, one that would allow for a more productive use of the Committee and ensure a louder voice for the Federation in Washington. In the Board meeting, a question was raised as to the financial impact the Committee's recommendations would have on the Federation's budget. Secretary Fuhrman indicated that the recommendations, if followed in their entirety, would require more than the funds available. In addition, he expressed the opinion that the procedures had, indeed, been followed and that contacts with the Congress and appropriate administrative departments had been productive, and further that the aims and objectives of the Federation were not consistent with those of a full-time lobbying organization. Despite these concerns, the recommendations were adopted by the Board at the Thursday meeting.

Legislative Seminars

Another important topic that surfaced at the 1966 Board of Control meeting and was approved for the following spring was the first in a long line of legislative seminars, initially sponsored jointly by the Federation and the American Water Works Association. After 1970, sponsorship was solely by WPCF. The initial seminar and those that followed provided a forum for the discussion of and an exchange of ideas on timely issues facing everyone involved in the water pollution control field. Over the years, these seminars became an important aspect of the Federation's government affairs program.

In 1967 and 1968, the work of the Legislative Committee continued with the review of numerous bills and assistance with regard to the presentation of Federation testimony before congressional committees. More importantly, the debate over how to improve the effectiveness of the Federation's legislative activity continued. At the 1967 Board meeting, for example, Mr. Smallhorst, in the absence of Chairman Vondrick, questioned whether the Federation had enough influence over legislation and policies and called for improvement of the reports of the Legislative Committee. With regard to intra-Federation relationships, the Legislative Committee suggested that it be represented in Federation activities such as conferences with legislators con-

One of the sessions at the Federation's March 29, 1977 Government Affairs Seminar in Washington, D.C.

cerning the subject of water pollution control legislation and regulation. Two Washington-based members familiar with the legislative process were appointed to the Committee.

In 1968, the Board took action in opposition to a proposed amendment to PL 89-753 that would have given federal agencies the authority to veto permits issued by state or interstate agencies for waste discharges into interstate or navigable waters, thus placing the authority completely in the hands of the federal agencies. Inasmuch as the proposed amendment conflicted with Point 4 of the Federation's Statement of Policy, the Board voted unanimously to transmit the Federation's opposition to this proposed amendment to appropriate members of Congress.

Committee name changed

At the 1969 Board of Control meeting, the Legislative Committee was made a constitutional committee and named the Government Affairs Committee to reflect more accurately the Committee charge. As set forth in Section 8.9 of the Federation's Constitution, the duties of the committee are enumerated as including the review and analysis of federal legislation in water pollution control and related fields, assistance to the Federation officers and staff in matters related to federal activity in this area, and organization and conduct of legislative seminars. In addition, the Committee is charged with "cooperation with other organizations engaged in similar or allied activities."

Another matter addressed at the 1969 Board meeting was the Committee's effort in support of Congressman Dingell's efforts to appropriate $1 billion for treatment plant construction grants for fiscal year 1970. On short notice, the Board approved sending a night letter, drafted by Victor G. Wagner, Government Affairs Committee Chairman, to all members of the House of Representatives urging a $1 billion authorization on the grounds that failure to do so would nullify previous local and state efforts to abate water pollution.

In 1969, the Board also approved two resolutions that focused on specific problems plaguing the national program. Because extreme concern existed regarding the inability of local governments to finance vitally needed abatement projects if the tax-free status of municipal bonds were lost, the Federation resolved to urge the Congress to oppose any legislation that would alter the existing tax-exempt status of municipal bonds. Traditionally such tax-free status had permitted local governments to finance public improvements at reasonable cost.

Furthermore, citing the confusion generated by having several federal agencies with variable procedures administering the program, the Federation resolved to urge the Congress to take necessary action to consolidate in a single federal agency all grant-in-aid programs for the construction of wastewater collection and treatment facilities for the abatement of water pollution in the United States.

Enter EPA

Within the year, the Environmental Protection Agency was created through President Nixon's Reorganization Plan No. 3 of 1970. Executive Secretary Canham indicated at the 1970 Board meeting that centralization of environmental responsibility might well improve federal action, but such responsibility should be expanded to include all government grant programs scattered among different departments rather than in the existing format of regulatory alignment.

At the same meeting, Government Affairs Chairman Wagner reported on the details of legislation passed, including PL 91-224, the Water Quality Act of 1970, the grant-in-aid program, and the recent successful legislative seminar. Mr. Wagner also cited the need for more Committee members and staff assistance in carrying out the Committee's program. During the 1970 meeting, the Board adopted a resolution urging the Congress to continue appropriations for research and development, pollution control programs, and construction programs for the abatement of pollution in view of their imminent expiration at the close of fiscal year 1971.

At the 1971 Board of Control meeting, Chairman Wagner presented the report of the Government Affairs Committee highlighting the Committee's coordination of 15 presentations to congressional committees and EPA and noting the success of the 1971 Government Affairs Seminar. He also announced a special meeting to discuss EPA's proposal on eligibility of turn-key-constructed wastewater treatment projects so that the Federation would be in a position to respond to the Agency. Of particular note in 1971 was Mr. Wagner's recommendation concerning more involvement by all Member Associations in governmental affairs at the local and state levels and the formation of a sub-group to work on upcoming legislation on water quality.

PL 92-500

The over-riding concern of the 1972 Board of Control meeting was the imminent enactment of PL 92-500. Passed by both houses of Congress and before the President for consideration, the Board adopted a resolution urging the President to sign the bill. With respect to the structure and focus of the Government Affairs Committee, Chairman Wagner reported that the Committee had considered more direct involvement on the part of the Federation with the Congress, the inclusion of a program on the relationships with EPA regional programs and the establishment of subcommittees and task groups to carry out an expanded government affairs program more effectively. In this regard, subcommittees were established on legislative development, government affairs, international, appropriations, and review and analysis. Task groups, moreover, were formed on state programs, state-federal relationships, research, coastal waters, Great Lakes, permits and monitoring, planning and land use, and operation and training.

Within a week after the annual Board meeting, on October 18, 1972, the Congress overrode the veto of the Federal Water Pollution Control Act Amendments of 1972 and the bill became law. These amendments greatly increased the level of federal funding for the construction of wastewater treatment facilities; expanded planning responsibilities at all levels of government; and established a regulatory mechanism requiring uniform technology-based effluent standards, together with a national permit system for all point-source discharges as a means of enforcement. These amendments gave the federal government final authority over most aspects of the program and moved it into a position to thoroughly dominate the field of water pollution control.

Of more importance was the implementation of this far-reaching and complex Act, which was expected to raise many questions from all segments of the water pollution control industry. Therefore, the Government Affairs Committee recommended, and the Executive Committee and Board of Control approved, in 1972, the concept of holding a pilot workshop on PL 92-500. The success of this workshop, held on December 12, 1972 in Romulus, Michigan, was immediately apparent and the Executive Committee authorized additional workshops to be held throughout the country in each of the 10 EPA regions. The workshops contributed to improved communication between municipal and industry systems managers, consultants, and federal and state officials concerned with the implementation requirements of the 1972 Amendments and eventually led to the adoption by the Board of Control on October 10, 1974 of the Government Affairs Committee-sponsored report titled "PL 92-500: Certain Recommendations of the Water Pollution Control Federation for Improving the Law and its Administration." Following adoption of that report, more than 4,000 pamphlets were distributed to all Member Associations, to interested groups and individuals, to key officials in EPA and other federal agencies, and to pertinent Congressional committees and members. The response to the report, with minor exceptions, was excellent, with all of the recommendations being actively considered at various stages of the legislative process.

Another important step for the Federation's government affairs program was taken in 1973. Vic Wagner, who had presided during a five-year period of extraordinary growth of the Government Affairs Committee, was elected Vice-President of the Federation and relinquished the Committee chairmanship. E.J. Newbould was appointed Chairman. One of his first recommendations concerned the desirability of forming a close and continuing relationship with the National Commission on Water Quality created by section 315 of PL 92-500. Over the next three years, this relationship was accomplished with the Government Affairs Committee acting as the coordinating committee for the review of numerous NCWQ contractor studies and the massive NCWQ report, submitted to the Congress by the Commission early in 1975. Working closely with other Federation committees, the Government Affairs Committee was able to put together a useful and highly informative document which was presented by President Wagner before the Commission as a compilation of the WPCF view of the report.

WPCF Public Affairs Office

The Federation's government affairs program was strengthened in 1973 by the approval of the Board for a Public Affairs Office at the WPCF staff level. The purpose of this move was to assist the Assistant Executive Secretary in the government affairs area and to work closely with the Government Affairs Committee. Considering the complexities of the new law and the massive amount of paper work that it fostered, as well as the high degree of congressional interest in the program, the creation of the new office was timely.

In 1974, the Government Affairs Committee continued its work and began to look for ways to improve the effectiveness of the Committee. The expansion of the Committee, along with its organization into subcommittees and task forces, inspired a need for more formal recognition of the assigned duties of the members. Chairman Newbould therefore began to develop a

President Horace L. Smith, accompanied by Executive Secretary Robert A. Canham (left) and Vice President Martin Lang (right) testifying before the Subcommittee on Water Resources of the House Committee on Public Works and Transportation on March 1, 1977.

GAC organizational manual that contained statements of purpose for each of the GAC subcommittees and task forces. The manual was adopted by the Board of Control in 1975 and proved to be a useful device for reviewing membership on the committee and assuring that members are aware of their assigned duties.

In 1975, it became apparent that the Federation should become more actively involved in the Section 208 areawide waste management planning process. Chairman Newbould therefore recommended, and the Board of Control approved, the encouragement of member associations to hold Section 208 workshops, similar to those that were held on the implementation of PL 92-500 a few years before. Despite the apparent lack of interest, the Board in 1976 again discussed the subject and authorized a pilot 208 workshop to be held in the Central States region early in 1977.

After 1947, the Federation had strengthened its government affairs program to meet the increasing demands placed on its members through the continuous enactment of federal water pollution control legislation. As a result, the Federation became more and more active in the legislative and administrative arenas on behalf of its members. The success of the Annual Government

WPCF Executive Secretary Robert A. Canham was briefed on a wide range of domestic issues at a White House conference on July 27, 1971. He is shown, right foreground, being greeted by President Nixon. Top Administration officials and specialists conducted the comprehensive briefing.

Affairs Seminar attests to the strides the Federation made in this area. Similarily, the work of the Government Affairs Committee with regard to the review of regulatory and legislative material became more effective toward the end of the Federation's first half-century. Also important was the Committee's commitment to review various legislative proposals and to develop specific recommendations for the Congressional debate on amending PL 92-500. Membership on various EPA committees, such as the Section 208 Advisory Commission and the Technical Advisory Group (now Management Advisory Group) also increased the Federation's leverage in the administrative process. Finally, and perhaps most importantly, the Member Associations of the Federation became more involved over the years in the government affairs program, with the formation of approximately 20 Member Association Government Affairs Committees around the country, and more expected to be formed in the future.

Chapter Seven

Charter Association Histories

Seven sewage works organizations joined to form the Federation of Sewage Works Associations on October 16, 1928. Within less than a year, five more associations organized and became members of the Federation, and New Jersey, which had had a Sewage Works Association since 1916, created a "journal-receiving group" in 1929. It may be said, therefore, that there were 13 original members in the Federation because New Jersey was represented on the Board.

In preparing this Federation history, the Committee decided that the material in this chapter should cover the initial operations of those 13 charter Associations and present brief statements on each Association that joined the Federation after 1929. A list of the Federation's "charter" Member Associations is as follows:

Association	Founded	Joined Federation	Members*
Arizona Sewage Works	1927	1928	15
California Sewage Works	June 1928	1928	120
Central States Sewage Works†	December 1927	1928	62
Iowa Wastes Disposal Association	November 1927	1928	57
Maryland Water and Sewage Association	April 1927	1928	44
Pennsylvania Sewage Works Association	July 1926	1928	85
Texas Short School (Sewage)	January 1926	1928	28
Missouri Water and Sewage Conference	October 1925	1929	84
New England Sewage Works Association	April 1929	1929	96
New Jersey Sewage Works Conference Group	March 1929	1929	48
New York State Sewage Works Association	May 1929	1929	167
North Carolina Sewage Works Association	November 1923	1929	52
Oklahoma Waterworks Conference	March 1926	1929	21

* At time of affiliation
† Indiana, Wisconsin, and Illinois

Arizona

In early 1927, the Arizona Public Health Association held a conference in Prescott to discuss problems of public health, including water supply and sewage disposal. Although documentation of that meeting is scant, the meeting is generally considered to be the organizational meeting of what later became the Arizona Sewage and Water Works Association.

The second meeting of the Arizona Public Health and Sanitary Conference was held April 17-18, 1928 in Tucson. The report on that meeting follows:

> The program included papers on milk and water supply, and sewage disposal. The afternoon of June 17th was devoted to the presentation of five papers on sewage treatment followed by a trip to the Tucson sewage farm disposal plant.
>
> The papers on sewage disposal were as follows:
>
> 1. "Activated Sludge Plant at Grand Canyon," M.R. Tillotson, Superintendent, Grand Canyon National Park.
>
> 2. "Use of Chlorine in Sewage Treatment," A.L. Frick, Field Engineer, Wallace and Tiernan Company.
>
> 3. "Sewage Treatment Plants for Small Cities," Chester A. Smith, Burns and McDonnell, Los Angeles.
>
> 4. "The New Sewage Disposal Plant at Tucson," George Grove, City Engineer, Tucson.
>
> Discussion by F.M. Veatch, Kansas City, and A.M. Kivari, The Dorr Company.
>
> 5. "Broad Irrigation with Sewage," G.E.P. Smith, Irrigation Engineer, University of Arizona, Tucson. The papers are to be published in the July issue of the Quarterly Bulletin of the Arizona State Board of Health.
>
> There were eighty-five men and women in attendance. Informal luncheons were held each day at the University Commons. The formation of a sewage works association was discussed as a subsidiary of the Public Health Conference. Miss Jane H. Rider, Director of the State Laboratory, was elected to represent the association in the National Federation.

California

The following information appeared in Volume II, Number 1, of the *California Sewage Works Journal* in 1929. It was authored by Leon B. Reynolds, 1929 President of CSWA:

> As president of the California Sewage Works Association I wish to thank Commissioner Parker for his gracious welcome. The association is glad to come to Oakland for its second annual meeting. Workers interested in sewerage and sewage disposal

throughout the United States have been meeting in groups in connection with other kinds of associations for some years. . . . This federation was formed and its board of control held its first meeting in October, 1928, at which time the constitution and by-laws were adopted . . .

In June 1928 a group of fifty-one men interested in sanitation met to consider the propriety of organizing an association in California; it was voted unanimously so to do, an organizing committee was selected, and it was decided to hold the first annual meeting in conjunction with the League of California Municipalities one year ago this month. At that time our constitution was adopted, which reiterated as the object of our association the advancement of knowledge, in this case, 'through interchange between members of the association and others of information, experience, and opinions.'

This is the second annual meeting of the association. At the end of our first annual meeting one year ago we had 115 members, at the end of our spring meeting last March we had 181 members, and now our roster includes 207 names, the largest association in the federation. There are now at least fourteen associations joined together in the federation for mutual help, with several others under organization, and their total membership probably approximates one thousand.

I had the opportunity two months ago to attend the first annual inspection meeting of the New York State Sewage Works Association in Rochester, and this association now has a membership of 175, so that they may soon be passing us for the lead in numbers. S.E. Coburn of Boston, the president of the New England Association, was also present and we were called upon to tell about the work and progress of our associations.

It is no longer necessary to apologize for holding a public meeting of a sewage works association—they seem to be becoming the style all over the country. . . .

The first annual meeting of the Association was held in October 1928 and affiliation with the Federation was approved at that time.

Central States

On the silver anniversary meeting of the Central States Sewage Works Association, Walter A. Sperry, Superintendent of the Aurora Sanitary District, Aurora, Ill. wrote a 25-year history of the organization. Mr. Sperry provided the Federation History

Committee with a copy of that historical background. His letter of transmittal is included in the appendix because of its historical interest. The following material is extracted or quoted from that history.

So far as the records show, the idea of an association of sanitary engineers and sewage treatment plant operators seems to have originated in the offices of Pearse, Greeley and Hansen, Consulting Engineers, Chicago, Ill. Very likely the original suggestion may have been born in the fertile mind of Paul Hansen of that firm. However this may be, Paul Hansen was a guiding spirit in calling a meeting of interested parties on Saturday, December 17, 1927 at the Chicago University Club. This meeting was attended by the following; From Illinois: Langdon Pearse, Dr. F.W. Mohlman and Mr. Tolman of the Chicago Sanitary District; William E. Stanley, L.B. Turner, J.G. Melluish and Paul Hansen of the firm of Pearse, Greeley and Hansen, Chicago; W.M. Olson, Cook County Department of Health, Chicago; Dr. W.D. Hatfield, Superintendent, Decatur Sanitary District, Decatur; and Gus H. Radebaugh, Manager, Urbana-Champaign Sanitary District, Urbana. From Wisconsin: Robert Cramer, Chief Engineer, Sewerage Commission and N.W. Brauer, Superintendent, Sewage Treatment Works, Milwaukee; Prof. C.I. Corp, University of Wisconsin, Madison; Adolph Kanneberg, State Railroad Commission; L.F. Warrick, State Sanitary Engineer and James H. Mackin, Superintendent, Sewage Treatment Works, all of Madison. From Indiana: Lewis S. Finch, Director, Water and Sewage Departments, State Board of Health, Indianapolis.

It will be noted that Indiana, Wisconsin and Illinois were represented in this group. A letter dated December 1, 1927 signed by Mr. Samuel Greeley and addressed to Col. E.D. Rich, Director, Bureau of Engineering, State Department of Health, Lansing, Michigan suggested that Michigan join the Illinois, Indiana and Wisconsin group in the formation of an association. Michigan subsequently declined this invitation in favor of her own Michigan association.

The minutes of the organization meeting held December 17, 1927 are so fraught with historical interest and are so suggestive of the beginning birth pains of the proposed Federation of Sewage Works Associations that they are fully quoted as follows.

Following luncheon, Mr. Paul Hansen stated that the purpose of the meeting [was] to be a consideration of the formation of an organization of persons identified with the operation, design, construction and management of sewage works. He then called on those present to express their views which are as follows:

Mr. Robert Cramer stated his approval of the projected organization and visualized divisions of the United States which would be components of a national organization of those affiliated with sewage works.

Mr. Langdon Pearse told of the inception of the movement to organize the operators of sewage works, and stated that the publication of articles and papers which would be prepared by members of the proposed organization would be financed for a period of three years by contributions from certain proprietary companies. The American Public Health Association offered its mechanical plant for issuing publications. It was estimated that subscriptions would not reach over 500 probably during the 3-year period although ultimately it was estimated that 1,000 might be secured. Mr. Pearse favored a tri-state organization which later could be split up when the number of plant operators shall have increased sufficiently.

Cities in the United States of 10,000 or over contained about 300 operators only.

The committee of the American Water Works Association, numbering 25, of which Mr. Emerson and Mr. Pearse are members, said that about $5,000 per year have been pledged for financing the publication through the efforts of Mr. George W. Fuller.

Mr. Cramer expressed himself as favorable to making a start and would do his share in furthering the project; that it was a pioneering effort and as such would have to be carefully guided; and not too much [should be] expected for the first few years.

Mr. James Mackin spoke in favor of the project, and emphasized the necessity of uplifting the dignity of sewage plant operators and bettering the grade of plant operation. He favored a tri-state grouping of organization and also considered that the proposed organization would afford backing to the State Sanitary Engineers in the enforcement of proper operation of sewage plants. Mr. Adolph Kanneberg inquired, "Are we not starting at the wrong end? How is this proposed organization going to improve the status and effectiveness of sewage plant operation?"

Professor Corp stated in regard to the operation of small sewage plants that the first problem is to educate the public as well as the operators and this is a function beyond the province of this proposed Association and would be for other agencies, principally state agencies. He said that in connection with C.M. Baker, former State Sanitary Engineer of Wisconsin, that he had been attempting, for three years, to educate operators together with public and plant officials. The proposed organization would be a stimulus to local operators. Mr. G.H. Radebaugh attended last year the meeting of the New Jersey Sewage Works Association and was aroused to the great value of contacts and interchange of experience among sewage plant operators. Mr. Radebaugh thought most favorably of the proposed organization. It was then moved by Mr. Radebaugh and seconded by Mr. Mackin as follows: It is the purpose of this meeting that we form a Sewage Works Association. The motion on being put by the chairman was carried unanimously.

Dr. Wm. Hatfield spoke of having just attended a similar meeting at Lansing, Michigan at which were present representatives from sixteen sewage plants together with public officials and sanitary engineers. He believes that this proposed Association should include all groups and grades of active and prospective sewage plant representatives.

Mr. Lewis Finch said we cannot hope to expect Indiana's small plant operators to become active and to attend meetings of this organization under the prevailing conditions. He emphasized the possibility of interesting city engineers throughout Indiana to attend and emphasized the educational values of the proposed publication.

Mr. Warrick spoke of the great need of the exchange of experience and ideas relating to the operation of sewage works between states. About 212 municipalities in Wisconsin have some type of sewage disposal; 85 have regularly constituted and operated sewage plants; 8 of these serve populations in excess of 10,000. Probably 8 to 10 representatives of Wisconsin plants could be counted on to become members of the proposed Association; of the 85 probably 19 might be induced to attend meetings.

Mr. Warrick wanted to see an Association formed that would ultimately reach small town sewage plant operators; also separate state groupings of operators in addition to the proposed central Association.

The Chairman then said that this brings us to decide the following questions:
(1) Shall there be formed an interstate organization?
(2) What shall the characteristics of the membership of such an organization be?

Professor Corp moved as follows: That we proceed to the organization of a Central States Sewage Works Association. The motion was seconded by Mr. Brauer and on being put was carried unanimously.

At this point Mr. Langdon Pearse offered for consideration a proposed Constitution patterned after the New Jersey Sewage

Works Association. This draft was taken up article by article and appropriately amended to meet the ideas of those present. The New Jersey constitution, as amended, was subsequently adopted.

It was moved by Mr. Cramer that the Constitution as drafted and read be adopted. Upon being seconded by Mr. Warrick the motion was put and unanimously carried.

It was moved by Professor Corp that we proceed to the election of officers for the ensuing year. Upon being seconded by Mr. Warrick the motion was unanimously carried. Mr. Radebaugh moved the nomination of Mr. Cramer as President of the Central States Sewage Works Association. The nomination was seconded by Mr. Kanneberg. On motion the nominations were closed and the secretary told to cast the unanimous ballot in favor of Mr. Cramer. The following additional officers were duly elected: First Vice-President, Lewis S. Finch; Second Vice-President, Dr. F.W. Mohlman; Secretary, G.G. Radebaugh and Treasurer, L.F. Warrick.

President Cramer on taking the chair suggested the advisability of our having a delegate to the American Society of Civil Engineers which will meet in January in New York City.

Dr. Hatfield moved that Langdon Pearse being a member of the Sanitary Division of the American Society of Civil Engineers and to be in attendance at the New York meeting, be made the delegate of this Association which motion was unanimously carried.

It was suggested that the next meeting be held in Milwaukee, Wisconsin.

The motion to adjourn was carried.

A group of officers and members interested in the newly formed Central States Sewage Works Association met with the Engineering Society of Wisconsin at Madison on February 18, 1928, to organize committees and start a program of activities. Members in attendance were Robert Cramer, President; James Mackin, Frank Qumby, W.A. Pierce, E.J. Beatty, O.J. Muegge, E.J. Tully, and L.F. Warrick, all from Wisconsin.

President Cramer opened this meeting by stating the objectives of the Association to be as follows:

(1) To secure efficient operation of sewage treatment plants.

(2) To obtain adequate financial support for sewerage works, maintenance and operation.

(3) Affiliation with a national organization which could sponsor the publication of a sewage works journal.

Iowa

A narrative history of the Iowa Association is included in "Iowa's Heritage in Water Pollution Control," edited by Federation Past President (1963-64) Harris F. Seidel and published by the Iowa Water Pollution Control Association in 1974. The earliest operators' conference in Ames is described in an Iowa State College Bulletin as follows:

> In order that the importance of the proper operation of such plants might be brought to the attention of those in charge, a conference for sewage disposal plant operators was held at Iowa State College, Nov. 2 and 3, 1915. The program consisted of talks and demonstrations on the operation of sewage disposal plants, all explaining the necessity for better operation, and suggesting how the same might be secured. On the second afternoon inspections were made of a number of plants, which were found to be working with various degrees of efficiency, further emphasizing the importance of this subject. On this program the college enjoyed the co-operation of Mr. Lafayette Higgins, Engineer, Iowa State Board of Health, Mr. Langdon Pearse, Engineer in Charge of Sewage Disposal Investigations, Sanitary District of Chicago, Mr. C.P. Chase, Consulting Engineer; Mr. C.H. Currie, Consulting Engineer, and Mr. L.E. Rein, Pacific Flush Tank Co. Local men appearing on the program were Dean A. Marston, Prof. R.W. Crum, Prof. Max Levine and Mr. C.S. Nichols.

The complete registration list for this first conference includes what may now seem to be a surprising number of councilmen and other city officials. In those early years, however, it was not uncommon for a member of the city council to be assigned responsibility for a municipal department such as water or sewers and to actively manage that department with help from the town's labor force. Iowa already had approximately 70 waste treatment plants at this time.

It is of interest that the Iowa Section of the American Water Works Association was also organized in 1915 at a meeting in early December in Iowa City.

Several of the reports of the 1915 Conference indicated that the meetings would continue, but nothing happened until 1920, at least nothing with respect to organized activity for the disposal plant operators of Iowa; most of the engineering staff of the college, even those who remained on campus, were heavily involved in some aspect of World War I.

The Second Conference on the Operation of Sewage Disposal Plants was held at Iowa State College on December 9 and 10, 1920. This conference was similar in general character to the first conference held in 1915.

The registered attendance at this second conference was 40, with 29 Iowa municipalities represented. The concluding paragraph of the 1920 meeting report stated: "Much interest was shown by the delegates in the meetings of the Conference. The sentiment of those present seemed to favor yearly meetings of this character."

What were these early Iowa conferences like? The usual pattern consisted of two full days of sessions spread over three consecutive days. For example, a conference might begin after lunch on Wednesday and conclude at noon on Friday. These meetings were held on the Iowa State College campus at Ames, usually in Engineering (now Marston) Hall or the Chemistry Building. Each session was built around a formal program of brief talks or papers on specific subjects or treatment steps, with time allowed for questions and discussion. Short inspection trips were occasionally scheduled.

In addition to open discussion following each talk, operators were encouraged to bring up specific problems from their own plants for general discussion. The program consisted of lectures, demonstrations and roundtable discussions of the every-day problems encountered in the care and operation of sewage-treatment plants. Every effort was made to present a program of the greatest practical value to municipal officials and others directly or indirectly responsible for the satisfactory operation of such plants in Iowa. The talk and discussions making up the program were informal, and the various subjects treated in a nontechnical manner.

At several of the early conferences there was considerable emphasis on the nature and treatment of the industrial wastes which had already become very important in Iowa. There were the wastes from the meat packing, canning, dairy, and sugar beet industries. Max Levine was rapidly developing practical treatment methods for these wastes, and the college was attempting to interest the processors in attending these conferences, but with little response.

Registration at these early conferences ranged from 40 to 60. Those attending included plant operators, elected city officials, city engineers, consultants and the engineers from the State Department of Health. These conferences were held yearly from 1922 through 1927.

Iowans played a prominent role in founding the Federation as described in an earlier chapter. The Federation Implementing Committee organized in 1928 included five Iowans among its 38 members. They were: Clare Currie, Lindon Murphy, Max Levine, Jack Hinman, and Al Wieters.

During this time, steps were also being taken to organize an operators' association in Iowa which could become a part of the new Federation. This was done officially at the November 1927 meeting in Ames.

Following the Wednesday evening banquet Murphy proposed the discussion of forming an Iowa sewage works association. Harry Jenks, Clare Currie, Jack Hinman, Max Levine, and Prof. A.H. Fuller were asked, in turn, to comment on the possibilities and purposes of such an organization. Following their remarks the subject was thrown open for general discussion. The sentiment of those present was generally favorable.

The business meeting was reconvened on Thursday afternoon, November 10, and the Iowa Wastes Disposal Association was formally organized. It was not until 1945 that the name was changed to Iowa Sewage Works Association.

At the November 10 meeting the new constitution was approved and officers were elected for the first time. The officers were Clare Currie, President; Jack Hinman, Vice-President, Lindon J. Murphy, Secretary-Treasurer; and Hans Pedersen and M.E. Shade, Association Directors. Dues for the new association were established at $3.00 per year of which $2.00 was to be retained by the Iowa Association, and $1.00 was to be forwarded for Federation membership and a subscription to the forthcoming *Sewage Works Journal*.

Authorization was also given for representatives to take part in further meetings leading to the organization of the Federation. The two men selected were Jack Hinman and Max Levine, both already very active in this regard. They were also appointed as Iowa's two Directors on the first Board of Control.

The next meeting of the Iowa association at Ames in November 1928 was designated as the Tenth Annual Sewage Treatment Conference and Second Annual Meeting, Iowa Wastes Disposal Association. The record of that meeting includes the following:

> The usual order of business was followed; under the head of new business President Currie recommended that the Iowa Association formally affiliate with the Federation of Sewage Works Associations. After full discussion a motion was unanimously passed favoring affiliation.
>
> A motion was made and unanimously carried that the officers of the Iowa Association for the present year be retained for the ensuing year.

The Iowa Disposal Association, one of seven charter members when the Federation was formally organized in Chicago in October 1928, had a membership at the time of the Chicago meeting of 57, making it the fourth largest in the group.

Maryland

In the spring of 1927, Dr. Abel Wolman, then Chief Engineer of the Maryland State Department of Health, met with Harry R. Hall, Frank H. Dryden, Carl Hechmer, and Edward S. Hopkins for the purpose of discussing his proposal to create an organization of water and sewage plant operators in the state. This meeting resulted in the First Conference of Maryland Water and Sewage Plant Operators, held at the Engineers Club in Baltimore with an attendance of 154 on April 21 and 22, 1927.

At this meeting, a new organization was formed to be known as The Maryland Water and Sewerage Association with the following officers:

Abel Wolman	President
Frank H. Dryden	First Vice-President
Carl A. Hechmer	Second Vice-President
James V. Cannen	Treasurer
T.C. Schaetzle	Secretary

The objective of the Association, as covered in the Constitution at this meeting, was "the advancement of the knowledge of design, operation, construction, and management of water and sewerage systems and industrial wastes treatment works." It is of special interest that the papers presented at this first annual meeting were published in Volume 1, Number 3 of the Engineering Bulletins of Dr. Wolman's Department.

By the action taken at that meeting, Maryland became one of the first states to create a formal organization to sponsor the advancement of knowledge of plant operations, thus stressing the need for proper maintenance and adequate operation of water and sewerage facilities.

The first Executive Committee met several times prior to the 1928 meeting, considered matters of concern, and prepared a Constitution which listed membership as those persons in Maryland who were interested in any work concerning water supply, sewage, or waste treatment. It was suggested that the 1928 meeting be held in Baltimore in April and that future meetings be held in the central, eastern, and western portions of the state in successive years.

At the Second Annual Meeting on April 10 and 11, 1928, F.W. Dryden was elected President, Carl Hechmer and Albert Heard were elected Vice-Presidents, T.C. Schaetzle became Secretary, and Edward S. Hopkins became treasurer. After the adoption of the Constitution, Dr. Wolman reported that a Federation of Sewage Works Associations was being organized for the primary purpose of publishing articles pertaining to sewerage and sewage works and moved that the Maryland Association join this new organization when it was formally created. The motion was seconded by Mr. Dryden and passed.

The new Executive Committee met immediately after the 1928 annual meeting and created an Exhibit Committee for the primary purpose of raising funds with which to publish an annual proceedings that would contain the papers presented at the annual meetings. The 1928 proceedings were multigraphed and distributed, and became a printed document in 1929 which was continued for some years. Another important occurrence in 1928 was that people in Delaware were specifically invited to attend the 1929 meeting to be held in Salisbury, Md., only a few miles from the Delaware line. As a result of that meeting, the organization's area was expanded to cover two states and the name was changed to The Maryland-Delaware Water and Sewerage Assoication.

When the Committee of One Hundred was created in July 1927, under the Chairmanship of Charles A. Emerson, Jr., three members of the Maryland Association were committee members Abel Wolman, Robert M. Morse, and John H. Gregory. When the Federation was organized, 44 of the 128 members of the Maryland Association were members of the Federation.

In 1936, the name remained unchanged, but the Constitution was amended to allow residents of the District of Columbia to be Active Members of the Association. Until 1946 members of the Maryland Association were eligible to join the Federation but changes in the Maryland Association Constitution in that year made the association formal by creating a Sewage Works Section in the Maryland-Delaware Water and Sewerage Association. The association became the Chesapeake Water Pollution Control Association in 1968.

Pennsylvania

According to Charles H. Young, now retired from the Pennsylvania State Department of Health, "the idea for a meeting of sewage works people in Pennsylvania was first proposed in 1926 by Prof. Elton D. Walker, Head of the Civil Engineering Department of Pennsylvania State College. Prof. Walker discussed the idea with W.L. Stevenson, then Chief Engineer of the Pennsylvania Department of Health. Stevenson was enthusiastic. Prof. Walker mailed letters to persons he thought might be interested. Walker and Stevenson arranged the program and set the meeting date.

"The meeting was held July 13-14, 1926 in the Hydraulic Engineering Building. Thirty-eight persons attended and the program was good considering that it was 50 years ago. Among the attendees were those who were later to become well known in Federation work: W. Rudolfs, H.S. Hutton, E.D. Walker, C.H. Young, I.M. Glace, E. Sherman Chase (WPCF Pres. 1952-53), H.M. Freeburn, and C.L. Siebeit.

"Prof. Walker chaired the meeting and in the course of the proceedings raised the question of holding an annual meeting.

The reaction was positive and committees were appointed to work out details."

The May 1966 issue of the Pennsylvania Operators' Newsletter adds this information:

> On July 13, 14, 1926, a group interested in sewage treatment operation and the feasibility of sponsoring and organizing a group which had for its primary purpose, dissemination of knowledge to sewage works operators relative to the various phases of sewage treatment.
>
> A Committee was appointed to formulate a Constitution and by-laws for a permanent association and arrange for a meeting in 1927. This Committee consisted of the following: H.J. Baum, Chairman, Clarence R. Fox, D.M. Thompson, and Professor Elton D. Walker. Col. L.E. Burnside, C.A. Emerson and R. O'Donnell were also present.
>
> A year later a two-day meeting was held at State College July 14 and 15, 1927. A Constitution was adopted on July 14th. The following officers were elected at that time:
>
> > President: Prof. Elton D. Walker
> > 1st Vice President: H.E. Moses
> > 2nd Vice President: Col. L.E. Burnside
> > Secretary-Treasurer: R. O'Donnell
>
> The Pennsylvania Sewage Works Association affiliated with the Federation on October 16, 1928.

Texas

In 1966, W.S. Mahlie of the Ft. Worth Water Department wrote a history of the first 40 years of the Texas Association for publication in that year's Proceedings of the Association's Annual Conference. Mahlie's summary included the following paragraphs about the Texas Association that began as a "Sewage and Industrial Waste Section" of the Texas Short School:

> So far as is known no instructions or discussion on matters pertaining to sewage treatment were given in the short school prior to 1926.
>
> In that year at the 8th School in Fort Worth four papers on sewage appeared on the program. At the 9th School in 1927, thirteen papers were presented and at the 10th School in Houston in 1928, thirty-four papers were read. At this meeting, Dr. F.E. Gieseke, Director, Texas Engineering Experiment Station, outlined the sewage research work to be undertaken at A&M College.

A special committee presented recommendations for "a plan to assist in the formation of a National Society for sewerage works operators to study questions connected with sewage disposal." There were several recommendations made in the report, the first one of which read as follows: "That all members of this Association who are desirous of studying sewage questions be organized into a separate division to be known as the "Sewage Division of the Texas Section of the Southwest Water Works Association."

It is interesting to note at this time that on the original "Committee of One Hundred" appointed July 15, 1927, which organized the Sewage Works Federation, five of the committee were from Texas. They were Messrs. Ehlers, Morey, Fugate, Eggert and Mahlie, and that in the first issue of the Sewage Works Journal, there were articles by Cohen, Fugate and Stanley, and abstracts of the papers presented at the 1928 Short School. At the 11th School in Bryan, 1929, affiliation with the Federation of Sewage Works Associations was completed. A proposed constitution is shown on page 17 of the Proceedings of the Eleventh School. Officers were to be the Chairman, Vice Chairman, Secretary-Treasurer and Editor.

It appears that although a constitution was adopted at this meeting, no official action was taken toward electing officers, since the chairman of the sewage division is listed as a standing chairman of the sewage section from 1930 through 1936.

A list of the officers since the Sewage Section started shows that W.S. Mahlie was chairman and representative to the Federation Board for 1928-31.

According to various sources, the Texas Association was a 1928 member of the Federation but Mahlie's review says that the affiliation with the Federation was completed in 1929.

Missouri

The beginning of the Missouri Water & Sewerage Conference was at an organizational meeting held in 1925 on October 26 and 27 in Kirksville, Missouri. The group was then called the Missouri Conference on Water Purification. The chairman elected in 1925 was A.V. Graf, Chief Chemical Engineer of the Water Department of St. Louis.

Annual meetings were held after 1925, and at every meeting papers were given and discussions held on subjects relating to sewage. Even at the first meeting in 1925, there was a technical program which included sewage-treatment subjects. At the meeting in 1927 six papers were given relating to sewage. One of these was by Thomas Veatch, consulting engineer of Kansas City, on stream pollution.

The name of the group was changed at the fourth annual meeting held in Hannibal on November 15-17, 1928, to the Missouri Water & Sewerage Conference in order to become eligible for affiliation with the Federation of Sewage Works Associations, and the conference represented the Federation up until the formation of the Missouri Section of the Water Pollution Control Association.

New England

F. Wellington Gilcreas, who served the Federation as Program Chairman from 1941 through 1955, was one of the founders of the New England Sewage Works Association in April 1929 and served as its Secretary for nine years. There were 40 charter members of the New England Association who were Federation members. On the occasion of the silver anniversary of the New England Assocation, "Gil" Gilcreas wrote the story of the first nine years of NESWA "Highlights in the History of the New England Sewage Works Association." Parts of those highlights, as published in the NESWA Newsletter, are reproduced here:

> Shortly after the formation of the Federation of Sewage Works Associations in October 1928, and the concurrent publication of the first issue of the *Sewage Works Journal,* the idea of organizing in New England a group interested in sewage treatment to affiliate with the Federation developed. Thus at a meeting of the Sanitary Section of the Boston Society of Civil Engineers in December 1928, Stuart E. Coburn introduced a motion requesting that the Section appoint a special committee to investigate the feasibility of the Sanitary Section organizing within itself a separate sewage group to affiliate with the Federation. The chairman of the Sanitary Section, Mr. Ralph Horne, pursuant to this action, appointed as a special committee, Stuart E. Coburn, F. Wellington Gilcreas and Edward Wright with instructions to report at the next business meeting. This committee met and at the subsequent meeting of the Section, reported that after careful consideration of all the factors involved, it was considered impracticable for the Sanitary Section to affiliate as a sewage works organization with the Federation of Sewage Works Associations, but recommended that the Sanitary Section favor the development of a separate and new organization covering the entire New England area, to be known as the Sewage Works Association. This report was ac-

cepted and the committee discharged. The same committee, however, immediately reorganized itself as an independent group to proceed with plans for the development of the New England Sewage Works Association. The three original members of this committee invited Ralph W. Horne, Gordon M. Fair, Richard G. Tyler, Warren J. Scott of Connecticut and Julius W. Bugbee of Rhode Island to serve as an organizing group. Mr. Coburn was selected as chairman of this group and Mr. Gilcreas as secretary.

This group met several times to consider the plans and concluded the best interests of the profession would be served if an organization were developed to cover all of the New England area, which would join with other similar groups in the Federation of Sewage Works Associations. This would provide a professional association for operators and all others interested in the field of sewage treatment and would permit all the members to receive the new *Sewage Works Journal,* then a quarterly publication devoted to the problems of sewage and waste treatment.

Thus Messrs. Coburn and Gilcreas were instructed to write a circular letter to every sewage plant operator in New England and to all others interested in the profession of sewage treatment, asking if they would favor the development of the New England Sewage Works Association and inviting their attendance at a future meeting to organize such an association.

A suggested constitution was developed which would conform to the regulation of the Federation of Sewage Works Associations as a basis for affiliation with that larger organization.

Since no funds were available for clerical service or for postage, the offer of Metcalf & Eddy, Consulting Engineers, and Weston & Sampson, Consulting Engineers, to underwrite the clerical expenses preliminary to the organization of the association were most gratefully accepted and used.

The first major problem was the arrangement for a general meeting. It was suggested that at a scheduled session of the New England Health Institute in Hartford, Connecticut, the question of organizing the Sewage Works Association be made a part of the program agenda of the Sanitation Division of the Institute. The first meeting of what was to become the New England Sewage Works Association was thus held at the Hotel Bond in Hartford, April 23, 1929. There was a large, gratifying attendance of men from all over New England interested in development of this new association. As part of the program, Mr. Coburn outlined the purpose of the new association, the problems relative to its organization and affiliation with the Federation. Mr. Charles A. Emerson, President of the Federation, described the plans and programs of that group and the value of the new Journal. Complimentary copies of Volume 1—Number 1 of the *Sewage Works Journal* were distributed.

After some discussion, those present voted unanimously that the New England Sewage Works Association should be organized as of that date. The proposed constitution was read clause by clause and was unanimously adopted and thus the New England Sewage Works Association was officially formed. A slate of officers was proposed and the following were unanimously elected.

    President: Stuart E. Coburn
    1st Vice-President: Julius W. Bugbee
    2nd Vice-President: Warren J. Scott
    Treasurer: Roscoe H. Suttie
    Secretary: F. Wellington Gilcreas

Following the election of officers, the session adjourned for a short period to afford those in attendance an opportunity to sign as charter members. Forty individuals joined the new association, paying the dues for the first year of $2.00 per person. Of this amount, $1.00 was to be remitted to the Federation of Sewage Works Association as a subscription to the *Sewage Works Journal*. The remaining $1.00 was for the support of the infant association.

The first act of the association, following the election of its officers, was to unanimously request the officers to apply for admission of the New England Sewage Works Association as a member of the Federation of Sewage Works Associations. From this modest but auspicious beginning, the New England Sewage Works Association developed as an important member of the Federation and a well organized technical and professional association in the New England area.

The executive committee met for the first time in Boston on August 27, 1929 and adopted the distinctive official emblem of the new association. Plans for the first independent, one-day meeting of the new association were formulated. This meeting was held at Worcester, Massachusetts in October, at which time 69 members and guests were present for the program of prepared papers in the morning, luncheon at the Hotel Bancroft and round table discussions in the afternoon.

The history of the New England Association is replete with names of some of the giants of the first quarter century of the Federation's existence. In the list of the New England charter members were: Julius W. Bugbee, E. Sherman Chase, S.E. Coburn, Gordon M. Fair, F.W. Gilcreas, Ralph W. Horne, Paul F. Howard, Roy S. Lanphear, R.S. Rankin, Warren J. Scott, R.G. Tyler. Later famous members of the New England Assn. included Thomas Camp, Edward Cobb, Walter Shea, LeRoy Van Kleeck, and William S. Wise.

The first name of the organization, "The New England Conference Group," was changed in 1930 to "New England Sewage Works Association."

New Jersey

The New Jersey Sewage Works Association is often mentioned as having had some relation with the Federation starting in 1929. In truth, there was no real liaison. In 1928, New Jersey members preferred to be recognized as the "oldest Sewage Works Association in the United States," although in March 1929, a group within the New Jersey Association of persons willing to pay $1.00 to receive the Federation *Journal* was organized. That association continued until 1942. During those intervening years, the New Jersey Association did have an influential friend in Bill Orchard, who was from New Jersey and served as a Director-at-Large of the Federation and as the Federation's Finance Committee Chairman.

At the 25th annual meeting of the New Jersey Association in 1940, Orchard made another strong pitch to that Association in a paper titled "The Need for a Strong National Sewage Works Association." Two years later, New Jersey became a member of the Federation.

On the 50th anniversary of the founding of the New Jersey Sewage Works Association, at a meeting held in Atlantic City in 1965, Robert S. Shaw, then President-Elect of the Federation, wrote an excellent review of the past and present of the New Jersey Water Pollution Control Association. Extracts of that review follow:

> We as an association were conceived in the minds of three farsighted men, Dr. R.B. Fitz-Randolph, Clyde Potts and Chester G. Wigley—these men were the men in positions of responsibility in the administration of the public health program in the State of New Jersey who jointly conceived of our association.
>
> Early in January, 1916, Clyde Potts wrote to various municipalities and utility companies throughout the States "that had in operation, or were about to construct, modern sewage disposal plants" inviting each to send a representative to a meeting in Trenton "for the purpose of forming an association for the promotion of greater efficiency and economy in the operation of sewage systems and disposal plants." This meeting was held in the

State House, Trenton, New Jersey, as scheduled and its proceedings are recorded in "Proceedings of the Meeting of Sewage System Executives of the State of New Jersey held on January 28, 1916."

At this meeting the formal organization of the New Jersey Sewage Works Association was confirmed. A constitution was adopted and officers were elected. These officers were as follows: President, John R. Downes, Plainfield; Vice Presidents, Paul Molitor, Chatham, and I.Z. Collings, Collingswood Sewerage Company; Secretary-Treasurer, Frederick T. Parker, Atlantic City Sewage Company. There were 22 members of the association.

The constitution as adopted on January 28, 1916, included "the object of this association shall be the advancement of the knowledge of design, construction, operation, and management of sewage works and the encouragement, at an annual meeting of the members, of a friendly exchange of information and experience.

There were two other interesting pieces of business recorded in the minutes of that first annual meeting. The question was discussed as to whether men in charge of water works should be invited to become members of the association, but, in view of the fact that there are in existence in the state two water works associations and no sewerage associations, it was moved by Mr. Collings that this organization consist of sewerage men only and that it be called the New Jersey Sewage Works Association. This motion was duly carried.

Also of tremendous significance there was appointed at this first annual meeting a committee with instructions to report at the next meeting of the association as to the advisablity of applying for legislation which would require the licensing of "superintendents or operators" of sewage treatment plants in the state. At the annual meeting of the association in 1925 a special committee was appointed to look into the possibility of short courses being organized by the association, the State Department of Health and the State University (Rutgers). Professor Harry N. Lendall, chairman of that special committee, reported at the 1926 meeting:

"It seems to your committee that there is a possibility for this association to be of real service to the sewage plant operators and also to city officials and those connected with the operation and maintenance of sewer systems, disposal plants and works of that nature. In considering the subject along this line the committee has outlined somewhat of a tentative course, and that such a course to be of real value and not of a superficial nature should be from two to three weeks in length, running at least five and one-half days to the week and occupying all of the time of the ones attending such a course, and that the course itself should comprise a certain series of routine work based upon a text book and a certain amount of laboratory work throughout the period, and that also there should be outside lecturers who will speak on the various problems of sewage disposal, those lectures to be taken from outside of the State in order that a general scope of the work might be obtained also from men within the State so that local problems can be presented, and then also a certain number of inspection trips should be made to see what is actually being done within the State."

After considerable discussion, a motion was passed establishing a joint committee among the Association, the State University and the State department of health to investigate the establishment of a short course for operators.

The first short courses were held at Rutgers University in January 1927. Enrolled were a total of 19 men including two from Ohio, one from New York, Pennsylvania, Connecticut and Delaware and 13 from New Jersey. This course consisted of 16 hours of recitation, 23 hours of laboratory work and 12 hours of lectures and round table discussions. There were two inspection trips made, one to treatment plants in the southern part of the State and one to treatment plants in northern New Jersey. Short courses have been continued at Rutgers University and elsewhere in New Jersey ever since that first course in January of 1927. These courses have been revised and supplemented on many occasions.

In 1965, Mr. Shaw commented, "We did not affiliate with the Federation until 1942. How grossly incongruous!" Bob Shaw, however, went on to say: . . . "it would appear that the non-member New Jersey Association did make substantial contributions of individual members during the intervening 12 years. It is interesting to note the registration of 450 at the New Jersey Silver Anniversary meeting in March of 1940.

"It was a short time after our Silver Anniversary meeting, highlighted by Bill Orchard's stirring address, that the New Jersey Association became affiliated with the Federation in 1942. At that time, Dr. Rudolfs, who had headed the New Jersey Conference group of some 60 members, proposed that his group be 'disbanded and the members who do not belong to the New Jersey Sewage Works Association become members and therefore help the New Jersey Sewage Works Association as well.' "

New York

Representatives from New York State expressed an interest in joining the Federation in 1928, but there was no formal organization in the state at that time. Representatives of New Jersey were also interested, but the New Jersey Sewage Works Association preferred not to join the Federation. The Federation's organizers decided, however, that both of these states should have some representation in the organization. Therefore, the bylaws were amended to provide directors-at-large. By this ploy, Bill Orchard and John Downes of New Jersey, W.W. Buffum and Kenneth Allen of New York, and H.W. Streeter of the U.S. Public Health Service in Cincinnati became directors.

A meeting was held in the Municipal Building, New York City on March 26, 1929 for "the purpose of organizing an association of persons interested in sewage treatment in New York State." Fourteen persons attended, including these men, who later became well known in the field and in the Federation:

C.L. Bogert (New York—Consulting Engineer)
Kenneth Allen (New York—Sanitary Engineer, Municipal)
William T. Carpenter (New York—Chemical, Municipal)
Earl Devendorf (Albany—Sanitation Engineer, State)
Morris Cohn (Schenectady—Sanitary Engineer, Municipal)
Richard H. Gould (New York—Sanitary Engineer, Municipal)
L.M. Fisher (New York—Sanitary Engineer, Federal)

The meeting was called to order by Mr. Allen, and by unanimous vote he was elected temporary chairman and Mr. Carpenter, temporary secretary.

The first order of business was the preparation of a constitution and bylaws and the chairman was asked to appoint a committee of himself and four more for this task. This was done but there is no record of who the other four were. Items discussed for inclusion in the constitution were: Should municipalities be associate or active members? Under what terms should persons residing outside the state become members?

The meeting did decide on a plan for furnishing leadership for the Association. The Executive Committee would consist of nine members with three elected annually to serve three years. The President and Vice President would be elected by the Executive Committee from its members and serve also as Chairman and Vice Chairman of the Executive Committee. The Executive Committee would also elect a Secretary-Treasurer, who would not be a member of the Committee. Consideration was also given to establishing a Sustaining Membership.

The meeting minutes also indicate that "Mr. Devendorf was elected Chairman of the Membership Committee with all present serving as members of that committee"; it was also voted that Albany would be the place for regular meetings.

The First Annual Meeting and Formal Organization Meeting of the New York State Sewage Works Association was held in Albany on May 4, 1929; 23 persons were present. These included 7 of the 14 present at the March 26 meeting. Minutes of that meeting describe these actions:

> Mr. Allen chaired this meeting and Mr. Bedell acted as temporary secretary in the absence of Mr. Carpenter. Mention was made of the various other local associations in the country and the Federation of those Associations. There was also a discussion on the need for such a local organization here and the benefits derived for such similar associations. The constitution and bylaws developed by the Committee on Constitution were adopted as read. Messrs. Cleveland, Suter, Cohn, Cox and Skinner were appointed as a nominating committee for the Executive Committee. They immediately went to work and produced the slate of Kenneth Allen, C.A. Holmquist, R. Suter, Morris Cohn, C.D. Holmes, Earl Devendorf, H.B. Cleveland, J.F. Skinner and R.H. Gould. There being no other nominations, it was voted unanimously to accept the report and declare the nominees elected. Mr. Allen reported that in response to the invitations sent out 84 persons had expressed a desire to become members of the organization. On motion of Mr. Shaw, seconded by Mr. Devendorf, it was voted that these persons together with any additional persons present (84 in all) be accepted as the original charter members upon payment of their dues.
>
> Referring to the work of the organization, Mr. Allen thought that one of the first problems of the Executive Committee would be to determine certain problems of sewage and waste disposal that should be studied and endeavor to secure the cooperation of laboratories or sewage works, both in municipalities and at the universities in the state.
>
> Mr. Devendorf stated that the State Dept. of Health had long recognized the need for stimulating interest in and having suitably trained operators for the sewage disposal works in the state and it would seem that this association would prove very helpful in securing this end.

Special attention was called to the fact that all persons applying for membership before July 1, 1929 would be considered charter members of the association. [There were 152 charter members.]

A recess was declared so the Executive Committee could organize. When the meeting resumed Mr. Allen reported that the Executive Committee had fixed on the following term of office for the various members: One year, Messrs. Allen, Gould and Suter; two years, Messrs. Cleveland, Cohn and Holmes; Three years, Messrs. Devendorf, Holmquist, and Skinner. Also, Mr. Allen was elected president, Mr. C.A. Holmquist, vice president, Mr. A.S. Bedell had been appointed secretary-treasurer, Mr. Allen to service on the Board of Control of the Federation of Sewage Works Associations for a one year term and Mr. C.A. Holmquist for the two year term. The next meeting would be at Rochester on July 27th and a third meeting in this year would be arranged for sometime in October, probably in Albany. The annual meeting was fixed to take place on the first Saturday after May 1st each year and would probably be held in Albany.

At the business meeting held on July 27, 1929, in Rochester, the Secretary reported there were now 167 members. A letter was read from the Secretary of the Federation of Sewage Works Associations dated July 8, 1929, stating that the Board of Control of the Federation had voted that the New York State Sewage Works Association be admitted to the Federation.

The President introduced Prof. L.B. Reynolds, president of the California Sewage Works Association, and S.E. Coburn, president of the New England Sewage Works Association, who presented greetings from their respective associations and gave very interesting accounts of the organization and work of their groups.

The technical session was a one-day affair consisting of four papers in the morning, a noon luncheon with the vice mayor as speaker and an inspection trip to the three sewage treatment plants in the city.

Pat Hill of the New York Association History Committee supplied the above information and added these comments:

The primary purpose, apparently, for organizing the New York State Sewage Works Association was to:

(1) Train operators through meetings at which the operators would exchange experiences relating to operational problems.

(2) Create a desire among operators and local officials for visits by knowledgeable personnel from the State Department of Health for on the job training of operators to assist them to improve operation and maintenance of treatment plants. A corollary to this was the hope that a spirit of friendship or camaraderie would be built up between operators and engineers from the Health Department so that the operators would call upon the engineers for help and advice when needed.

North Carolina

The first annual meeting of the North Carolina Water Purification and Sewage Treatment Conference was held in 1923. Reorganization as North Carolina Sewage Works Association was voted by the local executive committee in June 1928, and was approved at the sixth annual meeting of the Conference in November 1928, at which time Federation affiliation was also voted.

Oklahoma

According to the files of the Oklahoma Water Works Conference, it was officially organized on March 10, 1926 with 32 people present. The first constitution was adopted on November 17, 1926. Indications are that the first training conference was held in 1925, but no dates are available. The group met and organized at Oklahoma A & M College (now Oklahoma State University). M.E. Flentje was chairman and Dr. O.M. Smith was secretary of the organizational committee. The organizational officers were: M.E. Flentje, President; L.L. Smith, Vice President; and H.J. Darcey, Secretary-Treasurer.

Charter members were: L.L. Smith, Bert Sprangel, E.K. Skinner, W.H. Sutton, R.D. Morgan, H.M. LaRue, E.J. Crankshaw, R.W. Steinman, N.A. Kunkel, George Morris, W.E. Stockton, J.W. Heatley, E.L. Moon, V.A. Blackwood, J.R. Meigs, Ben McFall, J.T. McCarson, Arcel Love, H. Harris, W.L. Thompson, W.G. Curry, B. Cochran, R.C. Dohe, Robert Jones, W.I. Mathis, M.E. Flentje, Karl Bean, G.B. Savage, C.M. Means, R.E. Burke, S.H. McCurley, and H.J. Darcey

The records show that the association affiliated with the Federation on May 1, 1929

Other Member Associations

The order in which other Federation Member Associations joined the Federation is shown in the following listing with year of affiliation in parentheses.

Federal Water Quality Association (1930)

The Federal Sewage Research Association, now FWQA, covering USPHS employees, wherever stationed, was organized and affiliated with the Federation in 1930.

Michigan Water Pollution Control Association (1930)

First annual meeting of the Michigan Sewage Works Conference in 1925; at the sixth annual meeting in April, 1930, Federation affiliation was approved.

Ohio Water Pollution Control Conference (1932)

First annual meeting of the Ohio Sewage Works Conference in October, 1927; affiliation with the Federation was approved in 1932.

Institute of Water Pollution Control (United Kingdom) (1932)

In 1901 the Association of Managers of Sewage Disposal Works was founded in England; in 1902 it already had a membership of 90. The Association affiliated with the Federation in 1932. Later, it changed its name to the Institute of Sewage Purification and still later to the Institute of Water Pollution Control, as it is known today.

Institution of Public Health Engineers (United Kingdom) (1932)

The Institution of Sanitary Engineers affiliated in 1932. The Association changed its name to Institution of Public Health Engineers in 1956.

Canadian Institute on Sewage and Sanitation (1933)

The Institute dissolved in 1972 to allow organization of regional associations in Canada.

Kansas Water Pollution Control Association (1935)

First Operators Short School was held in April 1920, including both water and sewage subjects. Organization of Kansas Water Works Association at the annual short school in 1926; reorganization as Kansas Water and Sewage Works Association in about 1932; affiliation with the Federation approved in 1935.

Pacific Northwest Pollution Control Association (1935)

Organizational and first annual meeting of Pacific Northwest Sewage Works Association (Idaho, Oregon, Washington) in May 1935, in conjunction with the Northwest Conference on Stream Pollution; affiliation with the Federation approved 1935.

Georgia Water and Pollution Control Association (1936)

First annual Water School in 1932; second School and organizational meeting of Georgia Water Works Operators Association in 1933; broadened to Georgia Water and Sewage Association in 1935; Federation affiliation approved 1936.

Argentina Society of Engineers (1936)

Affiliation by the Sanitary Engineering Division of Argentina Society of Engineers occured in 1936; the affiliation was terminated in 1951.

North Dakota Water and Pollution Control Association (1936)

First annual meeting of North Dakota Water and Sewage Works Operators Conference in 1929. In 1936, both North Dakota and South Dakota affiliated with the Federation through the Dakota Water and Sewage Works Conference, which had separate state sections. In 1958, each state formed its own Association, with continuing Federation affiliation.

Rocky Mountain Water Pollution Control Association (1936)

First annual meeting of Rocky Mountain Water Works Association (Colorado, Wyoming, New Mexico) in 1926; first annual meeting of Rocky Mountain Sewage Works Association in 1936; Federation affiliation also approved in 1936.

South Dakota Water Pollution Control Association (1936)

First annual meeting of South Dakota Water and Sewage Works Operators Conference in 1935. (see also North Dakota)

Florida Pollution Control Association (1941)

First annual water works short school in 1931. Sewage subjects first included in 1936. Organization of Florida Sewage Works Association, and affiliation with the Federation in 1941.

Montana Water Pollution Control Association (1944)

First school for water works operators, including sewage subjects, in March 1932. Organizational meeting of Montana Sewage Works Association in May 1944; first annual meeting in April 1945.

Arkansas Water Pollution Control Association (1946)

First Water Works School in February 1931, which became Arkansas Water and Sewage Conference in 1935; Sewage Works Section for Federation affiliation formed in 1946.

Kentucky-Tennessee Water Pollution Control Association (1946)

Organizational meeting in the fall of 1946; affiliation voted by the Federation Board of Control in October 1946; first annual meeting of the Kentucky-Tennessee Industrial Wastes and Sewage Works Association in September 1947.

Virginia Water Pollution Control Association (1947)

First annual meeting of Virginia Water and Sewage Works Association in 1929, which by 1934 developed into the Virginia Section, AWWA. Organizational meeting of the Virginia Industrial Waste and Sewage Works Association in March 1947; first annual meeting in October 1947.

West Virginia Water Pollution Control Association (1947)

Organizational meeting of the West Virginia Sewage and Industrial Waste Association in June 1947; first annual meeting in October 1947.

Puerto Rico Water Pollution Control Association (1947)

The Puerto Rico Water and Sewage Works Association was founded in 1947, and also affiliated with the Federation in 1947. The name was changed to Puerto Rico Water Pollution Control Association in 1967.

Switzerland-Swiss Water Pollution Control Association (1947)

Affiliation by the Swiss Association of Water and Sewage Professionals in 1947.

Alabama Water and Pollution Control Association (1948)

First annual meeting, Alabama Water and Sewage Association in May 1946; creation of Sewage Works Section and affiliation with Federation in June 1948.

Louisiana Water Pollution Control Association (1949)

Organizational meeting of the Louisiana Conference on Water Supply and Sewerage in July 1938. Federation affiliation approved in October 1948, subject to creation in March 1949 of Sewage Works Section of the Louisiana Conference.

South Carolina Water and Pollution Control Association (1949)

Water Works Short School began in 1930's. The first sewage subjects were included in the Water Works Short School in 1945; establishment of the Sewage Section within the South Carolina Water and Sewage Works Association in 1949, with Federation affiliation.

Germany-Abwassertechnische Vereinigung e.V. (1950)

Affiliation by the German Sewage Technologists Association in 1950.

Nebraska Water Pollution Control Association (1952)

Organizational and first annual meeting of Nebraska Sewage and Industrial Wastes Association in April 1952.

Sweden-Swedish Association for Water Hygiene (1952)

Utah Water Pollution Control Association (1957)

First annual water and sewage works school in May 1950. Organizational and first annual meeting of Utah Sewage and Industrial Wastes Association in April 1957.

Mississippi Water Pollution Control Association (1957)

First annual short course for water and wastewater treatment plant operators was held at Mississippi State University in October 1955. Organizational meeting of the Mississippi Sewage and Industrial Wastes Association was held in August 1957, followed by Federation affiliation in October 1957 and the first annual meeting in March 1958.

Israel Association of Environmental Engineers (1957)

Affiliation by the Israel Association of Sewage Engineers in 1957.

New Zealand Water Supply and Disposal Association (1957)

Affiliation by the New Zealand Sewage and Industrial Wastes Association in 1957.

Indiana Water Pollution Control Association (1958)

Organized as the Indiana Sewage and Industrial Wastes Association, withdrew from Central States and affiliated with the Federation in November 1958.

Alaska Water Management Association (1960)

Hawaii Water Pollution Control Association (1962)

Organizational and first annual meeting of Hawaii Water Pollution Control Association in August 1962.

Australia Water and Wastewater Association (1962)

Nevada Water Pollution Control Association (1964)

Organization and affiliation of the Nevada Water Pollution Control Association in 1964 completed formal organization throughout the 50 states.

India-Indian Association for Water Pollution Control (1964)

The Netherlands-Nederlandse Vereniging voor Afvalwaterzuivering (1965)

Mexico-Sociedad Mexicana de Aguas (1965)

Japan Sewage Works Association (1966)

Venezuela-Asociacion Venezolana para el Control de la Polucion del Agua (1966)

Vietnam-Vietnamese Sanitary Engineers and Sanitation Association (1967)

Brazil-Associacao Brasileira de Ingenharia Sanitaria (1967)

Southern Africa Institute of Water Pollution Control (1968)

Italy-Italian Water Pollution Control Association (1969)

The Philippines-Philippine Water Pollution Control Association (1970)

Pollution Control Association of Ontario (1972)

Association Quebecoise des Techniques de l'Eau (1972)

Western Canada Pollution Control Association (1973)

Virgin Islands-Water Pollution Control Association of the Virgin Islands (1973)

Spain-Spanish Water Pollution Control Association (1974)

Finland-Vesiyhdistys r.y. (1974)

British Columbia Pollution Control Association (1976)

Korean Water Pollution Control Association (1976)

Chapter Eight

Personal Reminiscences

The following pages contain correspondence replying to the letters of inquiry mentioned in the Introduction. Also included are personal recollections of the seven Committee members, four of whom served as Federation Presidents and the three Executive Secretaries (one is also Federation Past President). Only the Chairman of the Committee has not served as a Federation officer, but he sat on the Board at annual conferences for about 16 years. This unusual circumstance is delineated in his recollections.

The inclusion of a limited and selected number of reminiscences was a decision of the History Committee Chairman, concurred in by a majority of the Committee. It was the original hope that reminiscences would be obtained from a dozen or so of the "founding fathers." Unfortunately, a number of persons asked to contribute failed to do so; many have since died. The chairman then decided to ask the Committee Members to contribute their recollections as sidelights to the history, and thus to obtain a good cross-section from 1940 on. Persons who are piqued because they were not asked to contribute may direct their ire at the Committee Chairman. It is hoped that with the publication of this history, other members of the Federation who recall interesting incidents in the early days of the organization will also submit their recollections for the archives.

George J. Schroepfer

Federation President, 1942-43

On January 18, 1940, the Board of Control of the then Federation of Sewage Works Associations held a dinner meeting in New York City. The meeting was quite informal and was attended by perhaps a dozen members. Principal topic of discussion centered around whether there should be a membership-at-large meeting of the Federation, including exhibits, rather than a business session of the Board as had been the practice since 1928. There was unanimous agreement that a general meeting of members was highly desirable, but there were major uncertainties—whether it should be held in 1940, and where was the best location. Considerable discussion revolved around various locations and the question of distance and convenience of travel.

The writer was present as the Central States Sewage Works Association's immediate past president and as incoming director. He sensed an unexpressed wish of the group that the meeting be held in the midwest part of the country, close to the center of membership of the Federation. Without having anyone with whom to discuss the possibility, the writer fell into the trap and extended an invitation on behalf of Central States to hold the meeting in that area. As a consequence, he was then and there appointed chairman of the General Convention Committee with the other members to be appointed at a later date. Having an invitation in hand, the Board of Control voted that the meeting should take place in early October of 1940. A later study of the membership dispersion developed that the center of membership was in Northern Indiana; as a result, Chicago was selected as the meeting place for the first general convention of the Federation.

Shortly thereafter, the full membership of the Committee was appointed by President Emerson; it consisted of the following subcommittee chairmen: Publicity—Anthony Anable, Advertising Manager of Dorr Co.; Exhibitors—O.T. Birkeness, Wallace and Tiernan Co.; Program—Cecil Calvert, Superintendent of Sewage Disposal, Indianapolis; Local Arrangements—F.W. Mohlman, Chief Chemist, Sanitary District of Chicago; Finance—William J. Orchard, General Manager, Wallace and Tiernan Co.; Entertainment—H.E. Schlenz, Pacific Flush Tank Co., Chicago. All preliminary arrangements were made by correspondence because there was no budget to cover travel and long distance telephone communications. A large number of men served on the various subcommittees. They were principally from the Central States Association, and chiefly from the Chicago area.

After considerable correspondence (the final total filled two file cabinets), a single meeting was held in Chicago in the spring of 1940, attended by the general convention committee and the key members of the local committees. After several hours, an impasse developed between the Chairman (Schroepfer) and Bill Orchard concerning the method of financing the meeting. With Dr. Mohlman acting as moderator and referee, the two retired to an adjoining room to attempt to resolve the problem. The manufacturers, represented by Mr. Orchard, desired to control the purse strings of the meeting for various reasons, possibly including the future direction that the Federation organization might take. The Chairman's position, supported by the Central States Association, was that no action be taken at this first meeting which would set a precedent of association or of direction of annual meetings and of the Federation itself.

In essence, it was hoped that this first meeting could stand on its own feet, with no finanical support and direction by the manufacturers, leaving to older and wiser heads (the Chairman had reached the ripe old age of 33) to set more deliberately the proper course of future procedure. Through the peace-making abilities of Dr. Mohlman, the problem was solved—or rather postponed for a year—by a verbal understanding that if a financial problem developed, the manufacturers would underwrite any loss to the extent of a maximum of $1,000. As it finally developed, the meeting was self-supporting, except to the extent of $300 which the Chairman wanted to pay out of his own pocket so as not to establish a Federation-manufacturers precedent; but was overruled by the other members of the Committee.

Another matter of concern to the Central States Sewage Works Association, and to others, was the effort that was being made by some of the manufacturers to effect a wedding of the Federation and the AWWA, which "hoped-for-plan" was always a spectre lurking in the background in the early years of the Federation. This continued until at least 1947 when a joint meeting of the Federation and the AWWA was held in San Francisco, which, incidentally, was the only meeting of the Federation which the Chairman has not attended.

The Chicago meeting on October 3-5, 1940 at the Sherman Hotel was a great success in all respects. The attendance was

556 out of a total membership of 2,819, or approximately 20 percent. Although this attendance does not match the usual present-day attendance of approximately one-third of the membership, it was considered a good showing for a first meeting.

There were 55 exhibits at the 1940 convention, which compares with several hundred at the present time. The exhibits, however, were of an entirely different character. They consisted of either four-foot-wide or eight-foot-wide spaces on backdrops where photos and literature were posted. Very few actual pieces of equipment were in evidence in the corridors of the mezzanine floor where the exhibits were held. Another difference was that the exhibits were generally not manned as they are today.

One similarity between then and now was the immense amount of volunteer time devoted by the membership to service on committees and on the Board. The dissimilarities between the 1940 meeting and the present meetings were largely in the financial and planning areas.

The budget for the entire meeting in 1940 was between $10,000 and $15,000 as contrasted with the present expenditure of several hundred thousand dollars. The time allowed for planning of the meeting also differed. Presently, convention cities are selected eight years in advance and planning begins two or three years before the convention is scheduled for a particular city. In the case of the Chicago meeting, less than nine months lead time was available with no previous pattern to follow. The planning started from scratch with everyone's expertise being limited to local meetings of one type or another.

The 1940 meeting demonstrated a great interest in and a real need for a national meeting. The establishment of the office of full-time secretary and editor and the appointment of W.H. Wisely to that position on January 15, 1941 launched the Federation on its way to becoming the great organization it is today.

The writer is a member of many professional organizations, but none has given him the enjoyment and satisfaction that association with the Federation has over the past 49 years.

Minneapolis, Minn.
Feb. 15, 1977

William H. Wisely

WPCF Executive Secretary-Editor, 1941-55

On October 3-5, 1940, I was attending the first national convention of the Federation in my capacities as engineer-manager of the Urbana and Champaign (Illinois) Sanitary District and as secretary of the Central States Sewage Works Association. I had undertaken the sanitary district job only six months before mainly for the opportunity to pursue graduate study at the University of Illinois with a view toward an ultimate career in research and teaching.

An invitation to the suite of William J. Orchard was accepted with no idea as to its portent. Arriving for the appointment, I found Mr. Orchard in the company of Charles A. Emerson and Arthur S. Bedell. They lost no time inquiring as to my possible interest in assuming the secretaryship of the Federation, if a suitable part-time arrangement could be worked out.

Suffice it to say that I gave up my original career plans and found myself in New York City on January 15, 1941, having that morning been appointed by the Board of Control as secretary of the Federation and managing editor of its journal. The reaction was just setting in. I was thinking, "What am I doing? Can I really handle all this the way it should be done and still meet my responsibilities to the sanitary district?"

I then knew AWWA Secretary Harry Jordan only slightly (by coincidence he and I had been born in the same tiny mining town of Coulterville in southern Illinois). Being near his office in midtown New York, I decided to pay him a visit. Gentleman that he was, he assuaged my uncertainties and propped up my confidence when I sorely needed a good pep talk. I returned to Champaign ready and willing to confront the future with the best I had to give.

In a general way, my 14 years with the Federation were an exciting kaleidoscope of hard work, long hours, and a host of the finest personal associations that could be available to any one man. My list of presidents included Charlie Emerson, Arthur Bedell, George Schroepfer, A M Rawn, Bert Berry, John Hoskins, Frank Friel, George Russell, V.M. Ehlers, Art Niles, Ralph Fuhrman, Earnest Boyce, Sherman Chase, Lou Fontenelli, and Dave Lee. Every one became a close, lifelong friend. One of my most cherished possessions is a certificate signed by them and betokening their "grateful and affectionate appreciation."

Lasting friendships were also formed with Floyd Mohlman, Ted Moses, Wellington Gilcreas, Willem Rudolfs, Langdon Pearse, Frank Jones, Morris Cohn, Bill Orchard, Linn Enslow, W.D. Hatfield, W.W. DeBerard, Harold Streeter, Abel Wolman, Gordon Fair and many, many more of the early pollution fighters. Adding in the list of friends made in the Member Associations in North America and Europe, the list becomes too long to comprehend. The fellowship and *esprit de corps* that existed throughout Federation officialdom and in the membership was to a great extent responsible for the success of the organization.

I owe a special tribute here to Professor Harold Babbitt, my teacher and firm friend, who was responsible in 1937 for my appointment as secretary of the Central States Sewage Works Association. From that time on my career moved inexorably into the field of engineering society administration.

Numerous references appear in this historical record of the Federation to the contributions of Dr. Floyd W. Mohlman and William J. Orchard. It was such a great personal privilege to work closely with both of these men that I feel compelled to make mention here of their extraordinary competence and character.

Floyd "Doc" Mohlman had no peer as a technical editor, in my experience. Despite his heavy responsibilities as Chief Chemist of the Sanitary District of Chicago, he devoted much of his professional energy for 16 years to the selection of papers and their preparation for publication in the *Journal*. His broad knowledge of the water pollution field enabled him to judge a manuscript with the fullest appreciation of its scientific or practical authority and value. The limited funds available for the *Journal* made necessary the optimum use of its editorial pages, and Doc never hesitated to exercise his judgment in the rejection of a paper regardless of the standing of the author. Wise, courageous, energetic and sincere, Dr. Mohlman was ever the gracious gentleman, helpful associate, and patient teacher.

Colorful, dynamic Bill Orchard—typical successful salesman—provided sagacious counsel and guidance on countless occasions as the Federation overcame the organizational and economic problems of its early years. As sales manager of Wallace and Tiernan, Inc., Bill always found time to serve the Federation wherever he could. His prime talent was the management of people and money, and the Federation was exceedingly fortunate in retaining his presence in the Board of Control as chairman of the Finance Committee for 35 years. He was an artist in drafting a budget. His astuteness in analyzing a business operation was phenomenal. Like Dr. Mohlman, Bill was always ready to share his knowledge and ideas with others, and he was never intolerant or unduly demanding. His sensitivity to the personalities and capabilities of his co-workers made him uniquely effective as an "organization man."

The approaching end of World War II brought an increasing amount of technical material and correspondence from Germany. Because translation service was not readily available, it seemed timely to undertake study of the German language. About this time Dr. Karl Imhoff sent me an autographed copy of the 14th edition of his *Tauschenbuch der Stadtentwasserung* and this was used as my basic study medium. It was a good choice, for Dr. Imhoff wrote most lucidly, using short sentences, unlike most German authors. After a few months of intensive study it was possible to read and translate all German letters and technical papers that came to my attention.

In July, 1954, it was my privilege to present Dr. Imhoff with his certificate of Honorary Membership in the Federation, in a ceremony staged at Bonn, Germany. A small part of the presentation was made in German, and the use of the *Tauschenbuch* as a language text was related. This was received with some amusement by Dr. Imhoff's associates, and especially by Mrs. Imhoff, who recalled several times in later years ". . . how Pete used Karl's book to learn German!"

The opportunity to meet and develop friendships with many other foreign notables in water pollution control was another valued reward in my work with the Federation. Among these were William T. Lockett, Arthur Key, C.B. Townend, S.H. Jenkins, R. Kershaw and Martin Lovett of England; N.J.N. Kessener, H.K. Baars, F.J. Ribbius and Ir Pasveer of Holland; Wilhelm Bucksteeg, Max Pruess, H. Muller-Neuhaus, Herbert Rohde and Franze Popel of Germany; Ed Holinger, Walter Dardel, Arnold Hoerler, Pierre Wildi and Karl Wuhrman of Switzerland; and Ingmar Gullstrom, Gunnar Akerlindh and Claus Fisherstrom of Sweden. Wonderful people, all! What a pleasure it was to know and work with them!

It seems that these reminiscences keep returning to the people with whom I was privileged to work in one way or another. My staff was never big in numbers, but it was great in terms of diligence, cheerful cooperation, and dedication.

Mrs. Gladys Roodhouse served well from 1941-43, when she and I had to share everything that needed doing. My personal secretaries later were Isabell Parnell and Mary Lou Hampton, and they were prizes, indeed. My first full-time editorial assistant was Mary Somers, who was capably succeeded by Mrs. Clothilde Skates.

It was a great day when we could afford an Assistant Editor, which post was filled in fine style for several years by Sylvan C. Martin. When "Sandy" moved on to a career in the Public Health Service, his place was taken by Herbert P. Orland, a top-flight, all-around publishing specialist.

By 1954 the budget even provided for an Executive Assistant, and Donald P. Schiesswohl was enticed from Florida to serve in this capaicty.

The accounting operation was handled competently by a number of bookkeepers, but John E. Hobart is especially remembered for his work in this area.

It seems fitting that the acknowledgement given the staff in my last report to the Board of Control be repeated here:

> My seven assistants in the headquarters office staff are deserving of the admiration and gratitude of every member of the Federation. Not only does this modest team conduct a full-fledged publishing operation—from advertising solicitation to circulation control—but also does it administer the affairs of a complex technical association engaged upon a host of activities encompassing hundreds of details. The progress of the Federation over the years is in large measure a direct result of the loyal, conscientious service that has always been forthcoming from my associates in the headquarters office.

Upon my return to London after that unforgettable meeting of the British ISP at Blackpool in June, 1954, a letter was waiting from a committee of the American Society of Civil Engineers, inquiring as to the possibility of my interest in appointment as Executive Secretary of that Society. This came like a bombshell, with all going so well in the Federation, and with my enthusiasm and personal relationships at their highest levels. My first reaction, after several days of deliberation, was to advise the committee that I could not reach a decision while in the midst of the exciting European trip, and that I realized that the committee could not await my return some weeks later.

The committee did wait, however, and invited me to a meeting for an interview. My dilemma resolved into a choice between work and personal associations which had completely enveloped my interests and professional aspirations on the one hand, and on the other a challenge to assume direction of one of the oldest and most prestigious engineering societies in the world. I could not resist that challenge, even though it meant sacrificing a job and a way of life that was fulfilling to the utmost. Now, having completed 17 years of service as Executive Director of ASCE, I know that my decision was the right one.

The difficulty of giving up the Federation work on January 15, 1955, was eased greatly by the selection of my competent and personable friend Ralph Fuhrman as my successor. The record speaks eloquently of his distinguished administration in the years to follow.

Gainesville, Fla.
Aug. 10, 1974

F. Wellington Gilcreas

Honorary Member, 1948

The project to develop a history of the Federation is most commendable and I am sure that a very valuable account will be produced. Any records that I had of the early days of the Federation were left in the Archives of the New England Association when my service as Secretary was terminated. Where they may be now is a question. I was not a member of the original Committee of One Hundred; Stuart Coburn represented New England on that body. After the meeting of that group, we started organizing the New England Association. Later I served on the early Federation Board of Control as the New England member. When Charlie Emerson suggested making the Federation into a true technical organization, rather than an editorial board, I was assigned to the Special Committee to plan and form such an organization.

Except for the New Jersey Association there was very little independent attention to sewage works and treatment science. Holmquist, Bedell, and Kenneth Allen were contemplating such technical development in New York State. In New England the Sanitary Section of the Boston Society of Civil Engineers functioned as the technical organization for the sewage treatment interests. In fact, that group was the original sponsor of the formation of the NESWA. I can perhaps assist in the report of that Association and its formation. My memory is still fairly good and perhaps I can give some accounts of the early operations of the Federation prior to 1940. Chester Brigham could probably help insofar as the New York Association is concerned as he worked closely with Bedell.

I wrote a history of the New England Association at the time of its 25th Anniversary in 1954. This covered the origin of that Association. I enclose a copy for your information and any use you may want to make of it. [See Chap. VII]. Recently this history was brought up to date and published in the new journal of the association. I think this was written by the Secretary at the time. I also enclose a list of the charter members of the Association, who joined on the organizing day of April 29, 1929. The names in brackets are those who had dropped out during the first 25 years. Since the date of this report, many have died but a goodly number are still active members, more or less like me.

My best recollections of the early days of the reorganization of the Federation are as follows. Prior to the development of the Federation as a national technical organization, there was a Board of Control with Charlie Emerson as the continuing President and Ted Moses as the Secretary. It consisted of one representative from each of the then existing local associations, together with Dr. Floyd Mohlman as editor of the *Journal* and representatives of the Chemical Foundation, who handled the funds from the Imhoff patents. Bedell represented the New York Association and usually I represented New England. The Board met once a year in New York during the month of January, usually at the Engineers' Club. The attendance of the local directors was generally scant and sporadic. Bill Orchard, Bill Piatt of North Carolina, Willem Rudolfs of New Jersey, Bedell of New York, and I were usually there. There were no funds for travel so those directors from a distance did not get there. Emerson, Moses, and Mohlman were always present.

The business, except for dinner, paid sometimes by the Chemical Foundation for those without expense accounts, was usually concerned with the problem of financing the *Journal*, its publication problems, how to expand advertising revenue, how to get more subscribers and organize more local groups. Incidentally, Wallace & Tiernan Co.'s contract to use the back cover of the *Journal* perpetually, did much to keep the *Journal* going.

At this time the New Jersey Association was not a part of the Federation since that group did not affiliate until 1942. Rudolfs had organized a small group in N.J. outside of the New Jersey

Association to be subscribers to the *Journal* and that group did belong to the Federation.

As the Imhoff funds dwindled, an active topic of discussion at these annual meetings was concerned with the then future and how the Federation and *Journal* could operate when these funds were exhausted. This prompted Charlie Emerson to present the idea of developing the Federation as a national technical organization, to remain also as a union of local associations, membership being by Associations rather than as individuals. Bill Orchard agreed to ask the Manufacturers Association to contribute $5,000 per year to the general operating funds of the *Journal* and the reorganized Federation. As you know, the Manufacturers agreed to do this.

Emerson then appointed a small committee to proceed with the details of such reorganization and in particular to plan and conduct a national meeting to be held in New York during October 1941. As far as I recall, this basic Committee consisted of Emerson as Chairman, Bedell, Orchard, Moses, and me.* In preparing for the meeting, Bedell was given the responsibility of local arrangements; Orchard, as you would anticipate, of finances, and I was given the program to plan and arrange. Morris Cohn and others whom I can't remember were of great assistance in the program problems. The meeting was held at the Statler in New York on October 9-11, 1941. The details of that meeting are part of the Federation archives.

The small committee was given other organizational duties, the details of which I don't remember except that Charlie Emerson and I were assigned the task of drafting a new constitution, which we did and which was adopted at the first meeting. Our original draft is unrecognizable now in comparison with the present document. We also prepared and wrote the first by-laws.

That about covers my immediate recollections of that period in the history of the Federation.

Regarding the early days of the New England Association, the Sanitary Section of the Boston Society of Civil Engineers appointed a special committee to consider the formation of a sewage works group and whether it should be a part of the Sec-

---

* Gordon Fair and Max Levine were also members of the Committee.

tion. The part played by this Society is covered in the History which I enclosed.

Regarding Federation programs, the original Constitution as adopted in 1941 provided for a Publications Committee to which was assigned the additional duty of preparing the program for the Annual Meeting of the Federation and assisting local Associations in program problems. This last was never a very onerous task, except as I helped Bedell with programs for the New York Association and as I assisted in the program developments of the N.E. Association, where I continued to be Secretary for a time. The Board of Control always appointed a Publications Committee, but it was always difficult to get much active work from the members, other than an occasional suggestion regarding a possible paper.

Funds for meetings of the Committee were never available. Thus as Chairman, I had to do most of the work and take or perhaps I assumed, the responsibility for planning programs. Thus the policy was developed of meeting with Floyd Mohlman on one or more occasions to discuss and plan the programs. The then President of the Federation was always asked to meet with us and many of them did. Funds for their travel were, of course, not provided at those times. So we all had to pay our expenses ourselves. This procedure continued during the early years of the Federation. Wisely as Executive Secretary was always consulted on program development although only by mail. Later, we did assemble the Program or Publications Committee when an opportunity presented itself. But Doc Mohlman and I usually completed and polished the programs for the annual meetings during our session in Buffalo. This system maintained the essential relationship between the editorial activities of the *Journal* and the meetings of the Federation. It was most effective and necessary during the years of war and the growth of the Federation.

The labors and contributions of Charlie Emerson and Bill Orchard during these early times cannot be overemphasized. These continued in the work of the Board of Control in the years just after 1941.

I hope that this long dissertation will be of help.

Gainesville, Fla.
Nov. 22, 1968

Earnest Boyce

Federation President, 1951-52

In answer to your query, I do remember some bits of information pertinent to the founding of the Federation. I did attend the dinner meeting held in Chicago in June, 1927, called at the request of Mr. Eddy, Sr. At this meeting, the earlier suggestion was made that AWWA (then meeting in Chicago) form a section on Sewage Works. I am enclosing a xerox copy of a report of this meeting [See Chap. II]. (I regret that my correspondence file of that period is not available to me, and I doubt that it still exists in the files of the Division of Sanitation, Kansas State Board of Health.) Several factors were stimulating interest in sewage treatment research and a need for a publication outlet was becoming evident. The formation of the Conference of State Sanitary Engineers in 1921 had more significance than perhaps we now appreciate.

The earlier annual meetings of the conference were held Saturday, Sunday, and Monday, just prior to the Annual Meeting of AWWA. This continued until sometime in the late 1930's, when it was changed to meet with APHA.

The need for instruction of water works personnel was a matter of great concern to the conference—and there developed during the 1920's a series of Water Works Conferences and Short Courses frequently jointly sponsored by a state university and by the sanitary engineering service of the State Health Department. Also, there were several state AWWA sections that were quite active. Texas was most active in the Southwest Water Works Association, and in cooperation with Vic Ehlers, started a water works short school perhaps prior to 1920. I know that as an assistant engineer for the Kansas State Board of

Health, I was sent to Waco, Texas, to attend a short school in 1923, and it was there that I met Vic. My assignment was to learn how the short course was organized. We held our first short course at the University of Kansas in 1924 and had Mr. Ehlers with us to help get started.

With the resignation of the Kansas Chief Sanitary Engineer in 1924 (Albert H. Jewell), I was appointed to his position and became responsible for continuing the Annual Water Works Schools held at the University of Kansas. The proceedings of these Schools were published in booklet form and I have the 8th and 9th proceedings of 1930 and 1931. The one for 1930 is Vol. 2, so judge that the first printed proceedings appeared in 1929. These were published by Kansas Water Works Association, which was formed in 1926 to "promote the best interests of the waterworks men in the state" and to "cooperate with the University of Kansas and the State Board of Health" in conducting the annual water works school.

It is of interest to note the names of persons who were invited to assist with these schools. Those assisting in February, 1930 included: Jack J. Hinman, Iowa City (and at that time President of AWWA; H.B. Crane of the International Filter Co.; Charles P. Hoover, Columbus, Ohio; N.T. Veatch, Jr. of Kansas City; and L.L. Hedgepeth, Penn Salt Co. While this was a water works group, each program had some paper on sewage treatment. For example, in 1930, H.W. Streeter gave a paper on "Sewage Treatment Practice in England," and C.A. Haskins, a consulting engineer in Kansas City, "Recent Developments in Sewage Disposal Practice." The next year, 1931, we had A.G. Fiedler, groundwater expert, USGS, W. Scott Johnson, Chief Engineer, Missouri State Health Department; W.C. Purdy, USPHS Station in Cincinnati; and R.E. Lawrence, then assistant engineer, Kansas State Board of Health. (Ray Lawrence was Federation President in 1961-62).

While I have recited in some detail the Kansas activities, similar programs were being carried forward in other states. Some states were organizing Water and/or Sewage Works Associations and there was a need to provide a publication outlet for the better papers. Our Kansas Transactions were from 140 to 150 pages and were financed through the sale of advertising (in the name of the Association).

I was not a member of the Committee of One Hundred but did have some correspondence with it. Following the June, 1927, meeting in Chicago, there were group meetings held in Cincinnati where the report of the Committee of One Hundred was given, and other meetings were held in connection with the APHA annual meeting in the fall of 1927 and at the annual meeting of ASCE in January, 1928.

As Secretary of the Kansas Water Works Association, I did represent the group in contact with those forming the Federation but it was not until about 1935 that we changed our Kansas Constitution to provide for a sewage works section which became affiliated with the Federation.

From that time on, I attended many of the annual meetings of the Board of Control of the Federation, representing Kansas until 1941 and then for some years as Chairman of the Federation Organization Committee.

Ann Arbor, Mich.
Oct. 16, 1968

E. Sherman Chase

Federation President, 1952-53

The following answers to your questions about the forming of the Federation are my most vivid recollections.

About organizations that existed before the Federation was organized, I remember:

The Boston Society of Civil Engineers organized its sanitary section in 1904; the American Society of Civil Engineers, Sanitary Engineering Division was authorized in 1922. Sometime prior to 1909 the American Public Health Association had an engineering section which dealt with sewage disposal. There were also a few local organizations of sewage plant operators, such as the New Jersey Sewage Plant Operators Association, which was organized in 1915-16. There was also an Ohio Association and one or two others. The first professional society that I joined (in 1909) was the American Public Health Association because of its sanitary engineering section.

With respect to the activities of the Committee of One Hundred that organized the Federation, I have very little recollection. According to my memory, the Metcalf & Eddy office had very little to do with that Committee and may not have had a member on it. I do seem to recall that Mr. Eddy, Sr. was a bit cool to the idea of another organization.

It is my recollection that the organization was set up as a federation of separate societies, each with a good deal of autonomy, instead of a monolithic organization like the American Water Works Association. This plan was adopted in order to overcome local objections to possible loss of identity. I can remember a good deal of opposition on the part of the New Jersey Association at the time Fischer Miller was president. Being the oldest organization dealing with sewage treatment, they felt pride of authorship.

When I was in Reading, Pa., Clarence Hoover, who was in charge of the then new treatment works at Columbus, Ohio, wrote around about 1910 to various people in the country suggesting the formation of an American Sewage Works Managers Association similar to the one then existing in England. Apparently there was very little interest at that time and the idea was dropped. At that time there were probably not over a half dozen technically trained men in charge of the operation of treatment plants, namely, Fales at Worcester, Bolling at Brockton, Bugbee at Providence, Hoover at Columbus, and Piatt at Winston-Salem, myself at Reading, and Lanphear in charge of the contact bed plant at Plainfield, N.J. in 1908 before John Downes. The Plainfield plant had been designed by Hering & Fuller. As a matter of fact, outside of small intermittent sand filter plants in New England and the middle west there were no plants treating the sewage from any community larger than, say, 200,000 population. Oh yes, there were one or two broad irrigation plants such as those at Framingham, Mass. and Pasadena, California.

In the mid-sixties, at the request of George Symons, Editor of *Water and Wastes Engineering*, I wrote a series of articles on the subject, "Nine Decades of Sanitary Engineering." From that series, the following paragraphs are pertinent to these "recollections."

> The Sanitary District of Chicago also initiated pioneering work in sewage disposal and treatment. Originally the sewage of Chicago was discharged directly into Lake Michigan which was also the source of water supply of the city. Healthwise this was tragic due to endemic water-borne typhoid fever. The Chicago Drainage Canal advised by Dr. Hering alleviated but did not solve the problem. Hence, the Sanitary District began its long series of investigations into methods of sewage treatment. Connected with these investigations from near their start was Langdon Pearse (1877-1956), whose direction thereof constitutes a notable achievement in the annals of sanitary engineering.
>
> Allied with Pearse was Dr. Floyd Mohlman as chief chemist. Mohlman was not only an able research chemist but was for many years the editor of The Sewage Works Journal, the official publication of the Federation of Sewage Works Associations. The demise of both Pearse and Mohlman is so recent that a description of their careers is hardly needed. Dr. Mohlman had studied under Dr. Edward Bartow, Chief of the Illinois State Water Survey, where early studies of the activated sludge process were carried on. Another graduate student with Dr. Mohlman at Illinois was Dr. W.D. Hatfield, a Federation past president.
>
> Reference to the Federation of Sewage Works Associations brings to mind Charles A. Emerson (1882-1955), who was not only an able engineer, but a key figure in the establishment of the Federation. From 1928, the date of its founding, until 1941, he was its president. Emerson graduated from Beloit College in 1903, and followed with a course in sanitary engineering at MIT, where like so many others he came under the influence of Sedgwick, graduating in 1905. His career included work at Columbus, Baltimore, with the Pennsylvania State Department of Health, and in consulting practice the latter years of his life.
>
> The first Sewage Works Association (whose half-century celebration will take place this year) in the United States was that formed in New Jersey in 1915. As I remember, among the pioneer spirits in its formation were Harry Croft, State Sanitary Engineer; Prof. Harry Lendell of Rutgers; John Downes of the Plainfield Sewage Treatment Plant; "Bill" Orchard of Wallace & Tiernan;

and S. Fischer Miller, President of the Pacific Flush Tank Co. There were undoubtedly others whose participation I have overlooked.

It is interesting to note that Charlie Emerson, who was so active in the formation of the Federation, was associated with Fuller at one time. It is also interesting that Emerson, Hyde, Pearse, Greeley, and many others came under the instruction and inspiration of Prof. Sedgwick at MIT when MIT had a course in sanitary engineering. Although Sedgwick was a biologist he was undoubtedly the father of modern sanitary engineering.

Further information as to early treatment facilities here in the U.S. appears in the book by M.N. Baker written probably 40 years ago and undoubtedly in the engineering library in New York. Most of the people who were active in sanitary engineering at the time of the formation of the Federation are now dead, except for Dick Gould and Louis Howson. Furthermore, Gordon Fair, and Abel Wolman who are only 10 or 12 years younger than I am, are still around.

This is a sort of rambling letter, but as I dictate I am thinking out loud and transmitting my mind back over its 60 years of activity in this field.

Auburndale, Mass.
Oct. 30, 1968

Gordon M. Fair

Honorary Member, 1955

One of these days when I get back to my office, I shall want to answer your letter of September 27, 1968 in which you write about a history of the Federation. Too bad that this was not thought of while Bill Orchard was still with us. His, after all, was the guiding hand in the whole matter.

What we really should do to assemble the information is for some of us old-timers to spend an evening together trotting out the yarns that we individually want to tell.*

One person whom I hope you will keep in mind as a source of information is F. Wellington Gilcreas, who was a very steady and powerful builder of the organization.

Unfortunately, it will be a few weeks before I do get to my office because, as you probably know, I broke my shin bone some seven weeks ago and do not expect to be fully mobile for about the same number of weeks in the future.

Cambridge, Mass.
October 7, 1968

Ralph E. Fuhrman

Federation President, 1950-51
Executive Secretary, 1955-69

My association with the Federation started on the first day that I went on duty as Assistant Public Health Engineer, Missouri State Board of Health in 1931. Herbert Bosch, my office mate and secretary of the State Water and Sewerage Conference, made it clear that a requirement of the job was Federation Membership at one dollar per year. Since that date, my membership has continued without missing a single *Journal*. The *Journal* was always a great help in my work.

Early in my association with the Federation, I became the Missouri Director on the Board of Control and a member of the

---

* Unfortunately, Prof. Fair's idea for a get-together of old-timers was not realized.

Membership Committee, chaired by Harold Streeter. All work was carried on by correspondence because there were no travel funds to attend the Annual Board Meetings held then at the New York meetings of the New York Sewage Works Association and the American Society of Civil Engineers each January. The principal duty of the Membership Committee was to review the documents of new member associations for conformance with Federation requirements, which could be done well by mail.

When I moved from Missouri to the District of Columbia Plant at the outbreak of the Second World War, I was named Chairman of the Federation War Service Committee. The proximity to New York made attendance of the January meetings both possible and stimulating. During the late 1930's the joint meeting of the Sanitary Engineering Division of the American Society of Civil Engineers and the New York State Sewage Works Association was the most influential and largest gathering of men in the wastewater treatment field in the United States. The organizations generously invited each other to join in sessions and they were often "standing room only," topping the attendance at any other ASCE session.

This setting lead to my first attendance of a Federation Board of Control meeting. I shall always remember the feeling of awe when I realized the principal participants were truly the "Who's Who" of water pollution control in the United States and Canada. Attending were Floyd Mohlman, Morris Cohn, Linn Enslow, Charlie Emerson, Bill Orchard, Wellington Gilcreas, Ted Moses, A M Rawn, Willem Rudolfs and Bert Berry, among others. In the early years thereafter, all of these greats and many others became close professional friends.

It was a natural development with such leadership and technical enthusiasm for the first Conference of the Federation as a trial to be set for Chicago in October, 1940. With the financial guarantee made by the Manufacturers Association, the Federation was reasonably assured of financial success which was realized and the pattern of Annual Conferences was established, with Board of Control meetings held with them. The Federation Conferences from that time on became the major annual gathering of men in the field.

The engagement of Pete Wisely as Secretary-Editor in 1941

soon proved to be a tremendous boon to the organization. As the first paid officer, initially for half-time and later full-time, he gave energy, leadership, and strength to the organization. It was at the right place at the right time.

It was truly a "bolt from the blue" when informed of my election to Vice President of the Federation at the 1949 Boston Conference. For the ensuing three years, it required detailed attention to Federation matters, particularly during my year as President. Sufficient time was never available, as I was also filling a full time job with extra work resulting from the design and construction of additional plant facilities and the enrollment at Johns Hopkins University for an advanced degree. However, with Pete Wisely's solid assistance the time passed with even an occasional visit to a member association meeting.

Through most of the 1940 decade, visits to such associations were made by the Secretary-Editor at Federation expense or the President without expense to the Federation. However, increased financial strength made possible support of officer travel to a much greater extent. This situation culminated in a 1954 visit to the European member associations by the Secretary-Editor. These contacts started a personal acquaintance with the European groups which has continued and strengthened ever since. In another decade, annual visits to virtually all member associations in the Western Hemisphere became regular events, along with frequent visits to others. In 1968, I was sent on a round-the-world trip including organizational contacts in Hawaii, New Zealand, Australia, India, Israel and the annual conference of the Institute of Water Pollution Control in Britain. In all cases the hosts were genuinely appreciative of the fact that the Federation was represented at their sessions, and the visiting officers had the reciprocal feeling.

By 1956, Conference registration became a vexing and time consuming event for all concerned. As a result, a plan of preregistration was devised along with a streamlined plan of handling money, effective with the 1957 Conference. Preregistration has been increasingly popular with more taking advantage of it every year since.

With the growing demand for Federation publicity and public relations activities but severely limited funds, the U.S. Public

Health Service, in 1957, made a news writer available to the Federation for publicity of the Conference in Boston. All went well except that the first Russian space shot, "Sputnik," was fired at the beginning of the Conference and our news releases were largely ignored. However, it made a good dry run, and in the following years the activity was continued with growing success.

Throughout my years as an employee of the Federation, I chose many employees for the varying tasks at hand. It is most gratifying that several of these people are still on the staff and are absorbed in engaging careers to their satisfaction and the benefit of the Federation. Of particular satisfaction is the service of Bob Canham who is already the longest-term employee of the Federation. His faithfulness and dedication are most noteworthy. All employees are deserving of commendation and thanks as much of the Federation success is directly attributable to them.

In closing, it should be recorded that many expressions of appreciation of the recepients of Federation honors they have received are made verbally in the excitement of Annual Conferences and are not recorded. However, there is one outstanding exception which is added here because so many in the field have had the benefits of his many attributes as Professor of Sanitary Engineering at Harvard University. His letter follows:

>Pierce Hall
>Cambridge, Mass. 02138
>October 15, 1964

Dear Ralph:

Now that the tumult of the annual meeting of the Federation is over, I bask in warm memories of gracious hosts, good friends, and generous associates. Mine was a noble and ennobling experience. I enjoyed it the more because I had a deep sense of belonging. Indeed, I felt that the proceedings of the meeting symbolized the rich experience of a lifetime.

For I well remember the founding of the Federation; its early years of small but steady accomplishment; the service of its journal as the cement between Sections at home and abroad; the encouragement of its friendly editor, Floyd Mohlman; the blunt but boosting judgments of its patron saint, Langdon Pearse; and the sound financial guidance of William Orchard. I could go on and on about remembered days.

By coincidence, my own career progressed in step with the evolving Federation—perhaps because of the Federation. This was not an isolated experience. How many others there were who marched with the Federation, I do not know; but I do know that no matter how many there were, they marched the better for the company.

It seems quite unnecessary to tell you that the Orchard Award means much to me, but say it I do with much affection.

Very sincerely yours,
Gordon M. Fair

It would be difficult to frame a more complete or appropriate expression of appreciation to this great organization.

Washington, D.C.
June, 1976

George E. Symons

Honorary Member, 1961

My introduction to the field of water pollution control came in the summer of 1925, when Dr. W.D. (Hap) Hatfield, employed me as a chemist in the sewage treatment plant of the Sanitary District of Decatur. Prior to that I had been working as the chemist of the Interstate Water Co. in Danville, Ill. At the time, I had completed two years of Chemical Engineering at the University of Wisconsin where I had had a course in water analysis.

"Hap" soon converted me from a clean water to a dirty water chemist and I began to run BOD's, the first of thousands upon thousands (I did both bachelor's and master's degree theses on BOD's).

It was Dr. Hatfield who sparked my ambition to continue my college career and to do graduate work. He was an alumnus of the Illinois State Water Survey (under Dr. Bartow) and he arranged with Dr. Buswell at the Survey to give me a job when I entered the University of Illinois in the fall of 1926, where I majored in sanitary chemistry and bacteriology, and was the undergraduate student chemist of the Illinois State Water Survey. In 1928 when the Federation was organized, I was in the last semester of my senior year at the University. The undergraduate engineer on the staff (who also graduated that year) was W.H. (Homer) Wisely, known as Pete during his Federation services as Executive Secretary, and as William H., when he became Executive Secretary of the American Society of Civil Engineers.

Dr. A.M. Buswell, Chief of the Illinois State Water Survey, literally sat out the formation of the Central States Sewage Works Association and the Federation. He would not join (and consequently none of his staff joined) because the Secretary selected for CSSWA was Gus H. Radebaugh, who was a prime mover in getting the Urbana-Champaign Sanitary District plant constructed. For Gus' enthusiasm and its results, he was picked as the first Superintendent of the treatment plant. His knowledge of sewage treatment was practically nonexistent. It was said, "He knew that it was needed and that was about all." Buswell was extremely annoyed that the District did not employ a professional (Wisely succeeded Radebaugh some 12 years later). Buzzie was doubly annoyed when Radebaugh was selected as Secretary of CSWWA. And so he sat out, although the Survey did subscribe to the *Journal* and the staff did contribute articles. But I didn't join until 1937, the year I went to the Buffalo Sewer Authority as Chief Chemist.

Shortly after I joined the NYSSWA (now NYWPCA), I helped organize and became first Secretary of the Western Section of that Association. Later I helped start the Metropolitan Section and contributed to the formation of the Hudson River Section. As I contributed more and more papers to the *Journal*—at one time I was third in number contributed after Rudolfs and Heukelekian—I went through the chairs of the New York Association, moving to New York in 1943 on my way to being President of that Association in 1945. That was a war year and the Association held no annual meeting—and as NYWPCA president, I had no meeting to preside over, but the several sections did hold meetings which I attended.

It was while in Buffalo that I became better acquainted with Floyd Mohlman, *Journal* Editor. At my suggestion, the National Aniline Co., one of the industries on the Buffalo River, called in Dr. Mohlman as consultant on the company's industrial waste problems. Mohlie (also an Illinois State Water Survey alumnus) came to Buffalo almost every Saturday. As F.W. Gilcreas tells in his personal recollections, it was there from 1940-43, or so, that Mohlie and Gil met and planned the annual Federation conference programs, picking the speakers, etc.

How different was that process from the one we developed when I became the first constitutional Program Committee Chairman in 1956-57. Gilcreas had kept a small three-ring black notebook of the programs he and Mohlman developed. He gave me that book and I used it as a guide for my five-year term. I passed it on to Paul Haney, who succeeded me as Program Chairman in 1962. It should be in the archives of the Federation.

As Program Chairman, I started the reporting of the number of hours devoted to each subject category. The last one I reported is included in Chapter VI. I recall that I picked Paul Haney as my vice chairman, and Larry Oeming as next in line. Each of them served as Chairman in succession after me. Ken Watson was Chairman of the Industrial Waste Committee and sat on the Program Committee. Ken was enthusistic about increasing the number of industrial papers on the program. To his ideas—not everyone agreed with him. I played devil's advocate and made him fight for everything that went on the program. He never failed to accomplish his ends. It was he who suggested Industry Day among other ideas.

In 1943, when I became Associate Editor of *Water & Sewage Works,* I began regular reporting coverage of the Federation Annual Conferences, and thus began a long period of sitting in Annual Board meetings as a proxy. In fact, Linn Enslow and I both were regularly given the proxy vote for Argentina and Germany. I think we did that for about eight years. Later I served as Director for the New York Association for three years and then as WPCF Program Chairman for five years. All together, then, I attended and voted at Board meetings for 16 years, or I believe, only six years less than Bill Orchard, Morris Cohn, F.W. Gilcreas and Willem Rudolfs, who served from 1940 through 1961 continuously in chairmanships that became subject to five-year limitations in 1956.

I remember some of those Board meetings were boring, some funny, some dramatic, some exciting. If one were to read the Board meeting minutes for many years, I think he would find that almost all of the motions were made by Bill Orchard and seconded by Morris Cohn. They always sat at the opposite ends of the head table where the officers sat. When I became a Director, Tony Fischer, Don Bloodgood, George Martin, and I cooked up a scheme to beat Bill and Morris to the making and seconding of motions. Between the four of us, we practically monopolized the motions and seconds for a couple of years.

Three of those four (Tony was not included in this scheme) arranged to spread our estimates of the Conference attendance—and we won the pot three years running, although the last year of the scheme we had a tie and split the total pot with someone long since forgotten.

I remember that as the years went by—and before the change in constitutional committee chairmen term limitation—there was more and more rumblings of the young Turks against the old guard running things forever. In one meeting, a Director, now deceased, said point blank to the late and great Dr. Willem Rudolfs, "Why don't you old bastards go away and die?" It was a tense moment, and the grief on Dr. Rudolf's face still lives with me. Eventually, the constitutional change came about, but not without some last-ditch verbal resistance from several of the Committee Chairmen who were to lose their seats and influence. That was a sad day, too, but essential to the growth of the Federation.

Other, more pleasant memories remain with me. Many of them were related to Annual Conferences. For the second and third annual meetings, I wrote and produced the *Daily Convention News,* reprints of which appear in the Appendix. Some years later, as a New Yorker, I was chosen to be Chairman of local arrangements of the Silver Anniversary Conference in New York City in 1952. In those days, the headquarter's staff did not do all the work that it does now. The local committee, and the

Arrangements Chairman, had a heavy responsibility. We hosted 1,152 registrants that year, including 209 ladies. Fifteen years later, in 1967, I was called on again to serve as Chairman of Local Arrangements for the 40th Anniversary Conference, also in New York City. It too, was a record breaker at the time.

Beginning in 1940, I attended every Federation annual conference. As I write this in the summer of 1976, I look back on the pleasure of appearing before the Federation Board of Control on many occasions, including once as the President of the American Water Works Association in 1973 and later as Chairman of the Federation History Committee.

Over the past 40 years, I have kept one foot on each side of the dividing line of clean and dirty water. I have enjoyed the many friendships, contacts, and acquaintances on both sides. I have been honored by both organizations and am happy to have had the opportunity to serve both. For my service as Program Chairman of the Federation, I was given Honorary Membership, and for my "service to the industry as a writer and editor," I received the Charles Alva Emerson Medal in 1962. One of my most lasting, closest, and most amusing friendships was with C.C. "Swede" Larson of Springfield, Ill., who served the Federation on numerous occasions, especially as the conductor of the Operator's Forum and as chairman of the Advertising Awards Committee.

The Bedell Award was a little harder to come by. Incidentally, it was I who introduced the resolution to change the name of the Kenneth Allen Award to the Bedell Award. About the Bedell Award I received, it was voted to me twice. The first was announced in the Executive Committee meeting of the NYWPCA, whereupon the then President of the Association turned to me and said, "George, would you mind if we gave the award to Chester this year; you can have it next time." Three, then six, then nine years went by before lightning struck me again.

Getting into the Quarter Century Club was ever harder. Woodbury Jones started that organization which was first officially known as the "Quarter Century Operators Club." The rules, however, limited membership to "men in resident charge" of treatment plant operations. I often argued with the late Morris Cohn and Henry Van der Vliet (each a one-time Registrar of the "Club") that it really was a quarter-century superintendent's club; ordinary operators, chemists, engineers, etc. could not qualify. Although I had first worked in a sewage treatment plant from 1936-43, I wasn't elected to the Club until about 1965. I was inducted into the Ted Moses Sludge Shovelers Society in the 1960's after I had conducted a "This is Your Life Ted Moses" show at a Pennsylvania meeting, the text of which appears in the Appendix. I received my other five shovels as the visiting AWWA officer at joint meetings.

During 37 years of attending Water Pollution Control Association meetings and conferences, I count 35 Federation meetings (there was none in 1945 and I missed 1976). During that time, I attended almost 200 Association meetings, at least once at practically every Association on this continent, perhaps 30 meetings of the New York Group, plus many in Canada, Central States, and other areas.

One of my most pleasant memories was writing the script for, and "M-C-ing" the "This is Your Life, Ted Moses" presentation in 1955. Another memory—serving as the editor of the 9th Edition of Standard Methods and working closely with Dr. Hatfield on the "Sewage Section" as well as contributing some analytical techniques. When I worked with A.M. Buswell in 1928-1933, I also edited the sixth edition of "Standard Methods," for which I received no money, only "Buzzie's" thanks.

Reminiscences usually are reserved for years well past, but I would feel remiss if I didn't make some comment about the period during which this History was written. It was in 1968 at Paul Haney's request that I undertook the task of chairing the committee to develop and edit the text for this History of the Federation. We expected that the work would be completed in four years. Neither Paul, Sid Berkowitz, and Ralph Fuhrman (who were in on the decision to do the history) nor I foresaw that Ralph Fuhrman would leave the Federation 18 months later, or that I would be elected vice president of AWWA in 1971. Both of these happenings had a profound and adverse affect on the production schedule of the History. Delays in contributions likewise added to the problem. The target date was moved from 1972 to 1974, then to 1976, and finally to 1977. So, what would have been a 45-year history became a history of the first half century of the Water Pollution Control Federation.

These past several years were not without their trials, tribulations, frustrations, and disagreements. Conversely, the cooperative efforts of the committee members were most satisfying. I shall not forget the experience and, on balance, I consider that it was a privilege to have served the Federation once more.

Larchmont, N.Y.
October 1976

Harris F. Seidel

Federation President, 1963-64

In the spring of 1962, I was working in Honduras as a short-term consultant with the U.S. AID program, helping develop a large water supply project for possible funding by one of the international banks. On arriving at my local home one evening from a strenuous field trip, I received a message that "a Mr. Johnson had called from Washington and would call back the next day." I couldn't remember anybody named Johnson in Washington, D.C. but hoped I had not committed some blunder or diplomatic incident which would put me on the next plane home.

The next day a call came but it was from Emil Jensen in Seattle. He was calling to say that I was the Nominating Committee's choice for Vice-President of the Federation. The connection was not the best, and it took a bit of doing for Emil to get this message across. A couple of days of reflection brought me to the conclusion that neither my family nor the City of Ames would be at all receptive to my beginning on such an adventure. With that, I put it out of mind until returning to Iowa about a month later.

Then the members of the Nominating Committee and Ralph Fuhrman began to call and write their encouragement. To my surprise, my wife Katy and the city "fathers" were willing to have me go ahead. Only then did I begin to consider it seriously. But it was weeks later before I accepted.

It has long been a tradition in the Federation that the office seeks out the man rather than the reverse, but surely few candidates have had to be pursued to this extent.

I didn't have long to prepare for my term as Vice-President which began in October 1962. I had never served on the Executive Committee, which was almost routine for those moving into the officers' chairs. However, I had represented Iowa on the Board of Control for the 1956-59 term and had also served as proxy for others several times, which provided some background in Federation affairs.

I was particularly fortunate in being able to draw on the experience and judgment of three Presidents who had served before me. Jack McKee made a special effort to bring me into the mainstream of Federation affairs as rapidly and as fully as possible. Ray Lawrence gave me encouragement at some critical moments, and several times kept me from charging headlong into battles I was sure to lose.

Everyone who knew Harry Schlenz will remember how he served the Federation with head and heart during his years in office. That is, of course, only a half truth. Harry served the Federation from behind the scenes for many years both before and after that. It was most fitting that he should be the first chosen from the ranks of the manufacturers to head the organization. On several occasions I went to Chicago to spend part of a day with him in his office at Pacific Flush Tank Company for help and judgment on current problems.

It is no revelation that any group of officers must learn to work as a team. It was my great good fortune to be followed by Al Steffen who is now enjoying the groves of academe at Purdue University. One maverick in a line can stir up a storm but will probably conclude his term with many bruises and few results. Very few things are changed much in one year.

Al and I worked as a team right from the start. I only pressed for changes or causes about which he was equally enthusiastic. With his aggressive follow-through the next year, many of them were achieved. These included organizational changes and staff strengthening which gave the Federation a better chance of keeping Bob Canham, then editor, and Bob Rogers, who later moved up to the editor's position when Canham became Executive Secretary.

This seemed to be a time of severe growing pains. The problems of membership records, accounting, and some of the other service functions were growing just a bit faster than the Federation's capability of coping with these needs. The employment of Bob Dark as office manager in April 1964 brought to the Federation headquarters staff a new competence in these areas. Later, Bob took over management of the annual Federation conferences.

Every traveling officer counts the experiences and friendships made on the road as among his finest memories and indeed they are. The fine hospitality and all the other perquisites become ever easier to accept, and one begins to believe at least some of those flattering things said in the introductions. It is vitally necessary to try to retain one's perspective, and to remember that the transition back to real life is inevitable.

I recall some of the rather ceremonial one-day visits made by Federation officers to our Iowa Association meetings in the 1950s. Few people realized at the time that the visiting officer was often coming on his own time, sometimes even at his own expense. The Federation's budget for officer travel was extremely limited in those days. By the early 1960s, budget support was provided and the goal of having a Federation officer or staff member at each Member Association meeting, was essentially being met.

It was Harry Schlenz who encouraged me to prepare a written report of each visit describing the general character of the Association and its meeting, along with some general comments on leadership and strengths and weaknesses of the group. These reports went only to the Federation office and to the other officers as background briefing. Similar reports by other officers were of great value in my travels. I could represent the Federation far more effectively at a Member Association meeting when I knew as much as possible about the group beforehand.

In those days the wives of Federation officers seldom traveled with their husbands except on short trips which could be made by car. The Federation budget then had no provision for the wives, although they added so much. The local ladies, whether or not they had an organized program, no doubt felt that the presence of the national officer's wife was of far greater meaning and interest to them than that of the visiting fireman himself. Also, it had been standard practice for some time by AWWA to pay the wives' travel expenses for the national officers. In joint meetings and similar occasions, this was a plus for AWWA which was impossible to overlook.

Katy and I were able to make a number of meetings together in the family car. And family car it was, because we took along our two pre-schoolers on several of the longer trips. During this period, the Executive Committee voted to provide travel support for Federation officers' wives as well. Nancy and Al Steffen were the first to benefit.

Among our travel memories, several will always stand out. My first trip on behalf of the Federation was to the Texas meeting in March 1963. I had spent the previous fall and winter in the hospital and at home slowly recovering from hepatitis. The other officers had attended several meetings for me during the winter, but I was determined to go to Texas.

I left Iowa on a late afternoon flight. A series of flight delays cut that night's sleep to less than two hours on the trip down. The last leg of the trip the next morning was a hedge-hopping flight skirting an impending rainstorm. It was so rough that I lost my breakfast, and then felt progressively worse the rest of the way.

When the plane landed at College Station, I barely managed to stagger off and to make my unsteady way around a group of people at the foot of the steps who seemed to be waiting for some dignitary to appear. I made it to the waiting room, put my head down on my knees and conceded to myself that the doctor was right; I had no business being there.

When I got back into a normal sitting posture, that welcoming committee came over and began some cautious inquiries.

They were the Texas Association officers and were most gracious and understanding. I was delivered rather carefully to the A &M campus for the start of the annual short school, where fortunately, my first day responsibilities were only to get acquainted. A second breakfast and some more sleep put me in considerably better condition and the rest of the meeting was a fine experience in every way.

That evening the officers, including Pearl Goodwin, took me out to Snook for the traditional dinner and fellowship. What is Snook? No written description could do it justice. Let's just say Snook is a wide spot in the road featuring a combination bar-restaurant-billiard saloon like which there is no other—anywhere.

Another highlight was the Pennsylvania meeting of 1965. Joe Lagnese was then President of the Pennsylvania Association, which has a long tradition of strong programs and leadership. The business meeting was held as scheduled at 11:00 a.m. on the second day. Operator certification was a hot issue. Several attempts for mandatory certification had been narrowly defeated in the State Legislature, and a controversy developed over what to do next. What followed was charge and counter charge and the reopening of some old wounds. A lengthy floor battle was finally resolved with an informal agreement to add more operators to the Association's Certification Advisory Committee. Particularly remembered: the participation of Bill Sacra, John Yenchko, and Sam Zack in this discussion.

After routine business matters were taken care of, it was election time. The nominating committee presented what might be called the "establishment" slate for the coming year. To the surprise of at least half those in the packed house, an opposition slate was also nominated! This was no spur of the moment gesture. It was a well planned challenge and the opposition group had brought printed ballots to expedite matters.

Again, there was an extended give and take before the issue was resolved. By that time it was 2:30 p.m. The business meeting had gone straight through the lunch hour and very few people had left. Through it all, Joe Lagnese, one of the youngest men in the room, presided with poise and style. He couldn't have been as calm and cool as he appeared, but he kept control of both himself and the meeting. A remarkable performance.

There are a number of joint meetings of Federation Member Associations with AWWA Sections. In some cases, these are very congenial and a conscious effort is made to see they are well balanced. In other cases, the balance is not as good and the relationship may not be very comfortable. My own career interests and experience have been pretty well balanced between new and used water. I've been an officer of the Iowa Section AWWA and have been active in water utility activities and committees. When given the opportunity, I leaned hard on the side of joint effort and better understanding between water and wastewater interests.

One joint meeting was way out of balance. The hotel marquee listed only the AWWA Section meeting. The program gave scant space to the wastewater meetings which were held in a crowded, stuffy service room off the kitchen. By contrast, the water works group met in comfort in the air conditioned ballroom. Hardly anything was overlooked which could have been done to put down the Federation group, which had no hand in the choice of location or meeting arrangements. The wastewater group told me in effect, "We can't do anything about it. It's always been this way."

I went away outraged, and poured my feelings into a rather savage meeting report. That report apparently had wider circulation than intended. For the next year I heard about that meeting at odd times and places, and over the next several years there was quite an overhaul of the entire situation, thanks principally to national AWWA leadership.

My relations with the traveling AWWA officers during those years on the circuit were superb. I remember especially Sam Baxter who, of course, knows both the water and wastewater field very well and who could not be ungracious to anyone even if he tried. The others were John Copley of Elmira, N.Y., and Jerry Allen of Seattle, Wash. Neither, to my knowledge, had ever been involved with the wastewater field but both went out of their way to be friendly, cooperative, and helpful to me far beyond the call of normal professional courtesy.

One remembered incident occurred during the well balanced Kentucky-Tennessee meeting. Jerry Allen was called out of a meeting to talk to someone. The moment he realized that it was an interview by a local reporter, he asked the young man to wait a moment, came back into the meeting room, found me and insisted I join him for the interview. That is courtesy of a rare kind.

One last comment, this one on the Federation's Quarter Century Operators Club. Operation has been my primary interest since 1951. I had considered myself a small operator until John Copley finally got through to me with his often repeated statement that "there are no small operators; it's just that some of us operate facilities which are not as large as others."

Requirements for membership in the present Quarter Century Operators Club include neither operating nor working at it for 25 years. It is ironic that the requirements are instead that one must have been a plant superintendent, but for only one year, 25 years ago.

During my term as President, I started a movement to reform this Club. This was probably the least noticed and most short-lived reform movement of all time; there wasn't any encouragement at all.

Why can't we have a true 25-year operators club? One of the two would need a new name. I feel strongly that there is a need for this kind of recognition. There are not many organizations like this: no dues, no meetings, no committee assignments, just glory. Surely 25 years in operation (at any level) deserves at least that much in return!

Ames, Iowa
August 1976

Harold E. Babbitt

Honorary Member, 1964

To try to answer your inquiries is pleasant and easy. To reply helpfully is another matter—my memory for such details "ain't" what it used to be. My recollections are all of the Central States Sewage Works Association.

As I recall, Central States was composed of organizations representing Wisconsin, Indiana, Minnesota, and Illinois. We called our publication *The Windmill,* a combination of letters and abbreviations from the names of the states. G.H. (Gus) Radebaugh was the first Secretary. He had been a booster of sewage treatment in Champaign-Urbana, Illinois, and when the plant was built, he became the superintendent. His enthusiasm for water pollution control earned him the election although he was untrained in the field of sanitary engineering. Annual meetings of the Association were rotated among the four states.

I recall our pride in one of our local boys who made good—that was W.H. (Pete) Wisely, who went from the University to the Illinois Department of Health, then to Superintendent of the Champaign-Urbana Sanitary District, then to Executive Secretary of the Federation and finally to Executive Secretary of the American Society of Civil Engineers.

I can remember our Central States meeting in St. Paul back in 1951 when I was president of the Association. Some supplier had won a contract with the Minneapolis-St. Paul Sanitary District and put on a celebration that was really good. When one is going through the chairs of an organization one thinks he'll never forget the things that happen. But we do.

Seattle, Wash.
Oct. 10, 1968

Sidney A. Berkowitz

Federation President, 1967-68

A return to the past without a diary would be a confusing trip except for significant signposts to guide me along the road.

My first attendance at an Annual Federation Conference was in 1954 when Florida first hosted the meeting in Miami and hurricane rains welcomed the members, but it was the 1960 Conference in Philadelphia where I had the first opportunity to participate in even a small way. However, my late and very close friend and boss for many years, David Lee, Federation President 1954-55, was constantly involved with so many of the organization leaders that I was soon acquainted with many of them, even if only in a second-hand way.

In the early sixties, starting at the Board meeting in Philadelphia in 1960, perhaps by accident, I was briefly involved in assisting the late Morris Cohn in rewriting and editing, at the request of the Board, the first Federation Statement of Policy. I recall sitting with Morris Cohn, a master with words, on the mezzanine balcony of the headquarters hotel, as the rough draft developed by the Board was revised and rough edges smoothed.

Not long after, the Florida association named me to represent it on the Board as a Director. For two of my allotted three years, I was privileged to serve on the Executive Committee. During that period, covering Conferences from Toronto to Atlantic City, I had an intimate involvement with Federation management and problem areas. Over the years, it has been interesting to observe how frequently identical issues have concerned the Board. Not the least of this was organization structure and Board representation, especially as the Member Associations benefited from the increasing public awareness and governmental involvement.

As attendance at the Annual Conference increased and the demand for exhibit space likewise grew, a crisis situation was soon reached. It concerned the second Florida meeting in 1964 where there developed competition for common space to be occupied by exhibits and social events. If the exhibits could be struck early on Wednesday, the available banquet seating for that night could be substantially increased. Attempting to resolve the problem, I made a special trip to Toronto seeking to persuade the management of the Water and Wastewater Equipment Manufacturers Association, but even with support from several of their more influential members, the effort failed. As a consequence of that situation, future conferences would be required to have increasingly greater space available where competition between the several segments would be avoided without limiting either exhibits or social events. Incidentally, as it turned out, the worries for the banquet were minimized as a hurricane hit south Florida only days before the Conference, leaving visual evidence of the force of wind and rain as we gathered at Bal Harbour.

The really exciting, emotional, and significant memories began for me, as they must have for others, with that always unexpected telephone call from the late Past President Ray Lawrence. It was, of course, on advice that if my superiors would permit me the time necessary, the Nominating Committee would place my name before the Board to be Vice President. Dave Lee, who was fully aware of both the personal honor and prestige as well as the honor for the state, obtained approval from higher authority. That was the beginning of what were to be several memorable and rewarding years for both my wife, Frances, and myself. Frances was probably the first wife of an officer to attend all the member association meetings for a span of three years. Other wives before had attended some meetings, but since the Federation had not been in a position to defray all the expenses, the individual officers would have to pick up a major share and how often the spouse would go along was thus limited. The benefit to the Federation of having the visiting officer accompanied by his spouse proved to be invaluable.

There were two highlights of my year as President. An emotional one at the banquet in New York City when Art Caster passed the gavel to me and with several members of my family present, I attempted a few remarks on what the moment meant to me. The second highlight, in two separated events, was the opening session at Chicago over which I presided when, because of my religion, I should have been attending services in a synagogue for the New Year of my faith. I had wrestled with the problem of attending services and resolved it by doing what would come my way but once, whereas every other year I'd be at worship. And finally the banquet, where all Presidents must have much the same feelings as they are center stage for the entire evening while they present awards and confer honors on many individuals. The most special thrills for me were announcing the newly established award for excellence in engineering education in honor of Dr. Gordon Maskew Fair, under whom I had been privileged to study, and calling Past President Harry E. Schlenz to the platform to receive the William J. Orchard Medal for outstanding service to the Foundation.

As an aside to the pleasures of the meeting in Chicago, there was the stench left in the headquarters hotel by so many malcontents who had been present for the national political convention a short time prior to our arrival in the Windy City. But offsetting the bad taste and odor were such pleasurable events as having an elevator practically at my command, especially when guiding our special visitor Walter Reuther to the exhibits and up to the suite.

On the lighter side, but somehow more readily recalled, were such incidents as the earthquake which shook Santa Rosa, and me, when we were at the California meeting. And that very cold (−4 degrees) night in Hutchinson, Kansas as Paul and Nell Haney and Frances and I waited the arrival of George Symons who would speak to the Kansas meeting the next day. Or the arrival in San Juan after the New York meeting; as we posed for a photographer at the top of the stairs, a fellow passenger was heard to inquire as to the identity of the dignitaries who had arrived with her. Arriving at Billings Airport we were met on a very cold night by a large contingent, who then escorted us down the mountain with police sirens and lights flashing.

Finally what is recalled with great appreciation is the readily offered advice and counsel from so many individuals, including many who preceded or followed me in the highest office of the Federation. Ralph Fuhrman was always available and his great assistance smoothed the way in a manner which often made difficult decisions seem simple. As the years have passed since the late sixties, there are those many instances, particularly at the annual conferences, when people stop to say hello and recall with us the days when we had the pleasure to join in their member association meetings.

Jacksonville, Fla.
July 1975

Paul D. Haney

Federation President, 1968-69

The 1966 Federation Conference was held in my home city, Kansas City, Missouri, and I looked forward to it with feelings of pleasure enhanced, no doubt, by the knowledge that my 10-year stint on the Program Committee was just about over. Five of these 10 years had been a learning experience as vice-chairman under the able coaching of George Symons. The preparation of the annual program is, at times, a gruelling experience and always an intellectual challenge. It is axiomatic that anyone who enjoys work can have a wonderful time on the Program Committee.

As Program Committee chairman, I was privileged to sit with the Board and participate to a limited extent in Board meeting activities. I didn't realize then, that at a future time, I would participate far more actively. These Board sessions recalled my much earlier experiences as a Director, representing the Kansas Association. It was evident that, over the years, as the Federation grew, so did the number and complexity of Board actions required. The pace of the meetings today is not as leisurely as it once was. Never intended as solely an honorary post, the Director's job each year becomes more demanding, but also more interesting.

Returning again to early experiences as a Director, one couldn't help being impressed by the strong role so ably played by Pete Wisely during the Federation's early years. His personality, enthusiasm, and competence supplied much of the momentum that made the Federation go and grow.

In retrospect, I was most fortunate during my term as Program Committee chairman to inherit from George Symons a strong organization peopled with a number of knowledgeable and energetic working members, many of whom stayed with me for the duration. Memory dims with time and age, and I have neither the ability nor the space to name them all. These are a few: Floyd Byrd, Ken Watson, Art Caster, Hayse Black, Rick Ryckman, Pete Krenkel, Dewey Nicholson, Albert Ullrich, Harry Kramer, Glen Hopkins . . . Especially strong, enthusiastic support was given by Don Pierce and Larry Oeming, who served as vice-chairmen. Then, there was always strong staff support from Ralph Fuhrman, Bob Canham, and Bob Rogers. These never lacked for ideas and encouragement. Sincere thanks are also due Bonnie Sagl, who patiently drafted and redrafted the program outline as it developed form and substance. During this period, the early sixties, sessions multiplied and we made a major effort to catch up on that badly neglected field, so full of vexing problems, the collection system.

Came the 1966 Conference and I retired from the Program Committee along with a much appreciated Federation Service Award. My retirement was short lived. Not many months later, I received a call from Past-President Harry Schlenz asking if I would accept the nomination as WPCF Vice President. Following quick consultation with Tom Veatch, Tom Robinson, and Past-President Ray Lawrence, I accepted. This was the first of a series of events that tumbled over one another for the next three years and profoundly affected my professional life and the personal lives of both my wife, Nell, and me.

The Board named me Vice-President at the New York Conference in 1967, but my term of office was only a few minutes. It was at this meeting that the Constitution and By-Laws were amended providing for the office of President-Elect. For a time, I was neither Vice-President nor President-Elect. Finally, the parliamentary gears meshed, and it was agreed that I was, in fact, President-Elect.

Serving the Federation as a national officer is indeed an adventure. It all begins when one first enters the officer line-up and the pace quickens as the days, weeks, and months fly by. There is work to be done and your term as an officer is over in much too short a time. Time's inexorable march dismays you and suddenly you find yourself at that Conference where you, as

President, pass the gavel to your successor. Upon entering office, you may have held the impression that you knew the Federation. This stance is quickly abandoned when it is realized that your education has only started. There is committee activity and inactivity to be dealt with, sessions to plan and preside at, meetings to attend, decisions to be made, airplanes to catch, baggage to find, letters to write, people to meet, talk with, encourage, placate, and papers, always papers, to sort, read, file, staple, treasure, discard. The job could be a lonely one but isn't for there is help at hand from the other officers and the staff. I won't forget generous assistance from Sid Berkowitz, Frank Miller, Art Caster, Joe Hanlon, Joe Lagnese, Ralph Fuhrman, and Bob Canham. The hard-working Executive Committee also deserves far more thanks than I can give here.

The most stimulating activities of a Federation officer are the visits to Member Associations. Here, you and your wife learn the real meanings for the words "welcome" and "hospitality." It is at these meetings that you encounter the people, the events, the things from which enduring memories are built. The first visit inevitably leaves strong impressions. In my case, it was the fine Puerto Rico Association. The Member Association sponsored a reconvened session following the 1967 New York Conference. It was our first time at San Juan, a delightful city full of history and beauty. An informal organization known as the "Bull Bats" was formed there. This group still meets occasionally for the conduct of certain sober deliberations, but, as yet, has not seen fit to elect officers, levy dues, keep minutes, or appoint committees.

The San Juan meeting was followed in quick succession by Pacific Northwest and North Carolina. After that, the scenario becomes somewhat blurred but standout sessions of Member Associations were: Florida, New York, Michigan, California, Missouri, Kansas, Oklahoma, Louisiana, Pennsylvania, Arkansas, Ohio, Utah, Rocky Mountain, Indiana, Arizona. A ready recollection of the latter is Art Vondrick, impressively dressed in academic robes, presiding, as Influent Integrator, over the Select Society initiation ceremony. Sessions in Canada and Mexico were also great experiences. The Canadians generously invited me to assist them in toasting the Queen and the President of the U.S. I considered this a great privilege. The Mexican meeting, held in Guadalajara, Jalisco, offered the challenge of presenting a few remarks in Spanish. After sweating through this, it was pleasant to receive the members' sincere words of appreciation. At another Mexican meeting, I dislike recalling a flash bulb explosion (mine). No one was hurt. Everyone was frightened but I was forgiven this disturbance of a technical session.

Digressing briefly to the larger subject of the Federation's international image, efforts have been made to foster this and I hope they will be continued to the extent feasible. The Federation is, and should continue to be, a significant force in international affairs.

The Chicago Conference in 1968 marked the beginning of my year as President and there was a great deal of urgent business to be conducted at the Thursday morning Board meeting which went on until the afternoon. A poignant experience for my wife and me was the unexpected arrival of our son, Paul Alan, a few minutes before the honors banquet. Then a student at the University of Kansas, he interrupted a busy schedule and made a fast trip to Chicago to share with us the pleasure and excitement of the banquet. We shall always treasure this gesture of thoughtfulness and loyalty.

Following Chicago, there was no slackening in the onrush of events. In much too brief a time, another summer was at hand. With it, came the sure knowledge that the Dallas Conference was not far ahead. Things seemed to be going well, but then came an unexpected event which overshadowed all others— Ralph Fuhrman's decision to return to government service with the National Water Commission. Emergency actions were required, and as an interim measure, pending Board action, I named Bob Canham, Executive Secretary pro tempore. Later, it was my pleasure to assist in changing this to a more permanent title. He was reappointed in 1974. This was ample evidence of his ability and dedication and this action left me with a good feeling. In a large measure, the Federation is indebted to Ralph Fuhrman for sound, energetic leadership through difficult years of reorganization and substantial growth. Not the least of his contributions was the foresight to select and guide a worthy successor who was ready to pick up the reins when Ralph elected to change ranges.

The years of Ralph Fuhrman's service, 1955-69, were those in which I became active in Federation affairs and what a help it was to have him as a counselor and to count him and his charming, talented wife, Josephine, among my true friends. I was and still am inclined to refer to Ralph as "Mr. Federation" even though I realize others have just claim to a share of the title.

My presidential term ended at Dallas, and that city certainly abounds in fond memories. Even after Dallas, there were still things to be done: meetings to attend, paper work, policies to consider, but these are matters for others to reminisce about. Retirement still eludes me and I am currently enjoying working with the Federation as chairman of the Publications Committee.

More than enough has been said but one is expected to close with some sort of summary. It has been both a privilege and a pleasure to serve this fine organization made up of good people who have banded together to further something in which they believe. They are, by and large, a friendly, happy lot who take pride in their work and have fun doing it. It is good to be one of them. I became a Federation Life Member in 1974.

Kansas City, Mo.
August, 1976

L. Deacon Matter

Past President, Water Pollution
Control Association of Pennsylvania

Tonight* my purpose is to compare some of the earlier banquets of this Association with those of today. Tonight we have

---

* Excerpts from a talk at the annual Banquet of the Pennsylvania WPCA held at Pennsylvania State University in August 1969.

a large beautiful hall and more than 390 people; the president of the Federation is here; and there are awards to be made. When this Association came into existence, there was no large room, no banquet, no speaker, no ladies, and no Federation representative. We weren't even an Association, much less a member of the Federation, because there wasn't any Federation.

In 1926, there were few sewage treatment plants and fewer operators. As a result, we had poorly operated plants and polluted streams. Sewage was a bad word. The operator of a sewage plant was an outcast; people walked on the other side of the street when he came toward them. If he was a political appointee, he was appointed after the dog catcher. No funds were forthcoming to keep the plant clean, or to enlarge it, or even to operate it, in most cases. I know, because I was a Health Department man trying to get municipalities to spend money on plants.

Earlier I said, "Sewage was a bad word" . . . I remember attending one of the Sewage Works Federation meetings held in Boston. One night, coming down in the hotel elevator, I was wearing my badge when two elderly dowagers got on. One of them looked at my badge, sniffed and said, "Why do they have to call it sewage?" I could have told her, but I restrained myself. Some years later the Federation became more public relations conscious and changed its name to Water Pollution Control Federation. The politicians found out that the people were in favor of water pollution control; they all jumped on the band wagon and we never had any trouble getting funds thereafter. The state and federal government showered us with money—and so we were on our way to the present.

Now to understand our early banquets, we have to know why and how this Association was formed. This is the way it began. A few men from the State Health Department and the Department of Engineering of Pennsylvania State College and a few good sewage plant operators met here at the College to determine the best way to instruct the few sewage treatment plant operators we had. It was decided that an annual conference be held, at which papers would be presented by people who knew something about sewage and sewage treatment plants. It was also decided to invite operators and the people who had hired them to come here to the college for those annual conferences in order to learn more about sewage treatment.

It took a long time to grow from a membership of 10 to 100. In the beginning, when we came here for a two and a half day meeting, we slept in the dormitories, ate wherever we could find an open restaurant and held our meetings in the classroom. On hot days we met on the grass under the trees. At the end of each meeting we held what we called a banquet—at a restaurant. The meetings were stag, as were the banquets; no women allowed. After a time of banqueting at various restaurants, we arranged to hold the annual banquet at the Nittany Lion Inn—in the small dining room, known as the Penn State Room.

As we began, the Pennsylvania Association was a pauper. The dues were low; hardly enough to pay for postage and a few secretarial expenses. I can assure you that for the first ten years when I was secretary of the PSWA, I didn't receive one cent of salary. Times have changed.

The registration fees at the annual conference were used for conference expenses, and were intended to provide some money for the treasury—but by the time I had paid the conference bills and paid the college for damages to the dormitory, and paid the fines for those who got arrested, and bail for others, I went back to Wilkes-Barre with nothing for the treasury. All I could do was hope that someone would pay his dues early and that the State would give us a few postage stamps. Now, for comparison—today, we have 1,000 members and an annual budget of $25,000. My budget, at times, was less than zero.

In the early days, banquets were neither elaborate nor fancy. To ensure a full attendance, we had to have three things: good food, a speaker, and entertainment. Speakers were the easiest to obtain. We simply approached influential engineers in the country, put the heat on and got prominent engineers as our speakers. Of course they came at their own expense. Our treasury couldn't have afforded them. If you don't believe that we had great engineers, take a look at the list of men who were crowned with the High Hat by Ted Moses.

Food was so-so, and entertainment was amateurish, even though we paid for it. We had magicians, accordionists, humorists, soloists, duets, trios, and quartets—and, regularly, we had member Dave Evans to lead the group singing each year. Dave's voice always gave out in the middle of the banquet—an annual event, to which everyone looked forward.

When we didn't have any money, we had to improvise. One year Bernie Bush, my assistant, and I did a William Tell act, but even though he had a grapefruit on his head, I didn't shoot, but we had fun building the suspense for 15 minutes. We even got Robert Spurr Weston into the act that time.

From the beginning, Ted Moses acted as Master of Ceremonies and soon got into the habit of telling stories; some were from Reader's Digest, others in various shades of blue were from unknown sources. When the ladies began to attend the conferences and the banquet, I used to spend the afternoon of the banquet auditioning Ted's jokes to get rid of the worst ones. Early in our history, we started the High Hat Ceremony as a joke along with the Sludge Shoveler's Society. But the joke turned out to be a coveted honor—and the names inscribed in the book comprise a list of some of the most important and most honored men in the profession of sanitary engineering.

As I said, our early banquets were a far cry from the dignified ones we have today. Perhaps it's just as well, but we sure had some fun in those early days.

State College, Pa.
August, 1969

Robert A. Canham

Executive Secretary (1969-present)

My point of reminiscing begins in 1946 as a part of the large group of students who had been separated from military service following World War II and had returned to school. As a student it was in the early part of daily contact with Don Bloodgood that he made it clear that it was desirable to become a member of the Federation. After listening to him about the advantages and using the *Journal* as a reference I did become a member of the Central States Sewage Works Association in January of 1947.

That was a milestone because it began the period of interest and support of the organization although at that time I would not have predicted the degree of involvement that I would have later.

Because of the time period and the inability to attend more than one or two Central States meetings in the next 10 years I regret that I did not become acquainted with many of the real founders of the Federation. I recall very well one incident with Bill Orchard, however. Bill had a way of being involved in nearly everything that went on and during the period of discussions with Ralph Fuhrman about joining the staff Ralph asked me one day to have lunch with Bill and him. During the discussion with Bill he pulled out of his vest pocket several pencil stubs that were so short and blunt that it was difficult to visualize them as being of much use for writing. Bill made a special point of advising that one should always have at hand a pencil and should record anything and everything possible.

Bill's piano playing and singing and his quarter-flipping games at Board meetings were legend and I saw a few of these. It is interesting that the quarter games still persist at the Board lunches during the Annual Conferences.

When I joined the staff in 1957 I was the eighth employee, and some of the eight were part-time. In retrospect I wonder how all the things that were done actually did get done. The staff was so thin that the piles of work always seemed to get higher rather than lower. For several years I went nowhere (including the bathroom) without carrying a stack of manuscripts to read.

One of the turning points was in 1964 when the office moved from the small overcrowded quarters on Albemarle Street in Washington some four blocks to 3900 Wisconsin Avenue. With this move to a new building was the chance to lay out the quarters and thus improve the working conditions for the staff. It also provided an opportunity for a room to hold meetings in and the conference room proved to be most useful.

The mid and late 60's were the period of most rapid growth of the staff and it was during this period that the groundwork was laid for much broader services and activities. In fact the growth was so rapid that it was necessary to acquire additional space in the building at 3900 Wisconsin Avenue six or seven times in the 12 years there.

The 12 years with Ralph Fuhrman as his assistant were an unforgettable period and I cannot thank Ralph adequately for his guidance and support. It was a genuine pleasure, both professionally and personally, to work with Ralph.

With Ralph's sudden departure in 1969 a period of adjustment was necessary because there had been no opportunity to plan for the change. President Paul Haney and the Executive Committee were most cooperative and constructive and I shall ever be grateful for that support.

For some time it had been obvious that several things needed to be done and that the feeling of the membership and the Board was such that it was time to expand the activities and become involved in more things.

The first evident problem was that of financial stability. For some years the Federation had been quite thin on reserves and as the budget began rising sharply it was evident that some changes were necessary to obtain a financially sound operation. For example, the reserves of $138,000 at the end of 1968 were only $30,000 more than those at the end of 1960, whereas the budget had increased from $180,000 in 1961 to $800,000 in 1969. Obviously the financial planning needed substantial strengthening. It took four years to make the results of planning for increases in various sources of income begin to show while at the same time expanding outlays because of expanded programs, growth, and inflation.

However, during the fourth year of the financial planning the reserves began to increase substantially and during that time the Board of Control established a goal of reaching and maintaining 50 percent of the operating budget as reserves.

During the period of the early 1970's the Federation financial operation changed from a small, relatively easily controllable one to a more impersonal, complex one. The budget reached $1 million in 1972 and in 1977 is nearly $3 million. At the present time the Federation is in the early stages of another financial adjustment period wherein expenditures have caught up with income.

In the late 60's it was obvious that the exhibits portion of the Federation's Annual Conference had an enormous potential for short-term and long-term growth. Further, it was clear that the long-standing arrangement with the Water and Wastewater Equipment Manufacturers Association, whereby the latter arranged and managed the exhibits and made a voluntary donation to the Federation, was not good financial management and that the resulting equipment show was not necessarily representative of a full cross-section of equipment and materials.

As a result a two-year temporary arrangement was made whereby a joint committee managed the exhibits and the Federation had a part in the financial results. At the end of this period the Federation Board of Control decided that the Federation would assume the full control and management of the exhibits. The first year of the Federation exhibits was in 1974 in Denver. Immediately the number of exhibits grew substantially and since 1974 has nearly doubled. This has been done with two people on the staff.

Again, in retrospect tremendous changes have taken place in the last 10 years in many areas and if one were asked what the largest single influence was, the answer has to be the federal program expansion, particularly the increasing amounts of federal money being spent in the construction grants program for publicly owned treatment works.

Some of the senior Federation members remember well the milestone in 1956 when the amendments to the Federal Act first contained construction grant funds. The first year it was $50 million. The program was so bitterly opposed by the states that it was well into the second year before the first grant obligation was made. The only state representative at that time who supported the program was Dave Lee of Florida. It is likely that he had a much clearer understanding of what would happen in later years. How times have changed since those early days with the states now struggling to obtain what they consider to be their fair share of the subsidy money.

It was most interesting to observe and be a part of the evolution of the Federation's involvement in national affairs. Probably it was typical of a group of mostly professionals who wanted to have impact on the national program but who many times acted like the ball they were handling was just out of the fire. The process took quite a few years during which only a few seemed to have any real understanding of how such a group could have meaningful influence on the national program.

The commitment and results began to show nearly simultaneously when the Government Affairs Committee was given Constitutional status in October 1969 and in 1970 the staff was authorized to expand to provide support services for a continuing government affairs program. When Leo Weaver joined the staff in the early 1970's the program building began and during the three years of his guidance it became a viable and respected one.

The highest degree of involvement by the largest number of Federation members throughout the U.S., and the most satisfying results, was the series of workshops held during the consideration of what became PL 92-500. These workshops and the resulting documents probably did more to build the credibility of the Federation than any other single planned program before or since.

The Federation has a long way to go in order to reach the level of influence that my understanding of the Board's desire was in the initial period as well as in later years. It is a slow, and many times frustrating, process that requires patience and much persistence. I believe the right kind of progress has been made and hope that the commitment will continue, and further that it will continue to be based on the technical expertise that resides within the Federation in abundant quantities.

The past 20 years have been by far the most interesting, challenging, and satisfying period of my life. It has been deeply gratifying to be a part of the Federation's period of growth in size, stature, and maturity.

The close of the Federation's first 50 years certainly does represent a major milestone. The problems have grown in numbers and in complexity. The success in dealing with them can be appraised with mixed results, depending on the origin of the point of view. In general, though, the entire technical community can take credit for being a major part of the continuing effort to improve the quality of water and in turn the quality of life. The Federation has been a significant factor in that effort.

It is not correct to reminisce forward but if one were to be present at the end of the second 50 years and reminisced a bit my guess is that the techniques would have changed substantially but that there still would be much concern with not yet reaching the goal of clean water, not enough money to do everything, and that politics would be the same as they always have been.

Washington, D. C.
June 1977

Victor G. Wagner

Federation President, 1975-76

When the Federation History Committee decided to delay publication of the history until mid-1977 in order to have the volume published in the 50th anniversary year, it meant one thing—much of the final drafting of the text would take place in 1976 and 1977.

Therefore, when the Chairman of the Federation History Committee suggested that my Presidential Message published in the September 1976 *Journal* would, with some revisions in style, make an interesting and valuable addition to this chapter on

Reminiscences, I was happy to make the revisions suggested. These, then, are my comments and recollections.

In preparing for my Presidential Message, I reviewed the messages of the previous presidents, beginning with the man who wrote the first in 1962, Harry Schlenz. In each of the articles, there was an attempt to project the future based on an analysis of problems facing the industry at that particular time.

I was impressed with Harry's remarks, where he said that a newly-elected president, "impressed by the honor bestowed on him, finds that he makes promises of new and revolutionary programs overlooked by his predecessors, only to learn that . . . such new ideas had already been set and were merely ready to be implemented by the incumbent." We constantly refine what has been said, built on what has gone before, and implement new ideas of others. Truly, there is "nothing new under the sun," and my remarks are structured around the comments of my predecessors.

On Federation size, Jack McKee (1962-63) reviewed membership growth from 1930-1960. There was a three-fold growth every 15 years. Dr. McKee predicted a stabilized growth after 1960 to a membership of 15,000 in 1975, but by that year actual membership was about 24,000, still a three-fold rate every 15 years. Conference attendance has shown a four-fold growth about every 20 years or less.

These figures, WPCF's part in the expanding environmental movement, and the fact that only a small portion of people engaged in or associated with the pollution control industry were members of Member Associations, led me to the prediction that Federation membership will reach 70,000 by 1990, with a conference attendance of almost half that number in the same year.

McKee and others suggested the need to welcome other disciplines into the Federation: economists, lawyers, industrialists, physicists, agronomists, geologists, planners, limnologists, oceanographers, politicians and others interested in or associated with the environment. Creating an attitude in WPCF that will stimulate these individuals to join us, while we share with them knowledge and experience in dealing with environmental problems, will benefit all and ensure Federation growth.

It is fine to say that new disciplines should be welcomed as partners in WPCF, thereby increasing membership, but how to accomplish it? Involvement is the answer. This "new" idea has been around since the Federation began. Joe Lagnese (1971-72) spelled it out in his message, when he said, "Success of the Federation and its Member Associates depends significantly on the extent of individual member involvement."

In 1976 involvement was evident when the number of members serving on WPCF committees approached 1,000 and new committees were being formed, staffed, and challenged with new problems. More important was the involvement in member association activities which reflect credit to the entire membership.

Government affairs has been a prime example of the results of involvement. Nearly all past presidential messages expressed the need for increased activity in governmental affairs. The last three years have highlighted the positive effect of such involvement.

On October 18, 1972, the U.S. Congress passed the Water Pollution Control Act Amendments of 1972 (PL 92-500) over a presidential veto. The Federation's Board of Control anticipated the passage of this new law and the fact that this law would drastically change the concepts of water pollution control in the U.S. On October 12, 1972, the Board authorized a pilot workshop, to be co-sponsored with a Member Association, for the purpose of informing the membership about the law and assisting in the process of implementing it. This workshop was held in Detroit, Michigan in December 1972. The workshop was highly successful and other Member Associations immediately started planning similar workshops throughout the country.

In Floyd Byrd's 1973 message, "The Federation, A Dynamic Organization," he emphasized that these workshops provided a means for bringing together representatives of EPA, state, and local governments, consultants and industrialists for an interchange of ideas and opinions.

John Parkhurst in 1974 referred to these workshops in identifying problems that had developed in implementing PL 92-500. The Federation Board of Control reacted to these problems by approving the document, "PL 92-500, Certain Recommendations of the Water Pollution Control Federation for Improving the Law and Its Administration." This document provided the basis for the Federation's presentations to the National Commission on Water Quality and testimony before committees of Congress. It also served as reference material for action by many member associations. With the establishment of the National Commission on Water Quality under PL 92-500, the Federation's active interest in government affairs was continued. On April 6, 1976, the Federation held its 10th Annual Government Affairs Seminar, developed around the Commission Report. The theme was "Public Law 92-500—Mid-Course Correction." The program included an address by a member of the Commission, an analysis of the report by members of the staff that worked on the report, comments by EPA representatives and staff members of the Senate Public Works Committee and the House Committee on Public Works and Transportation.

One of the early WPCF Presidents who commented on the need for greater involvement in government affairs was Art Caster in 1967. Art reported that, at the request of Frank C. DiLuzio, Assistant Secretary of Interior, a Federation Committee was established to confer at intervals with the Assistant Secretary regarding water pollution control problems. The members of that Committee were Caster, Harry Schlenz, Ralph Fuhrman, George E. Symons, Donald J. O'Conner, J. Floyd Byrd, Emil C. Jensen, Charles B. Kaiser Jr., and Ray Lawrence. Caster forecast that the Federation would be called upon for more involvement in the future.

President Art Vondrick in 1971 pursued the idea of involvement in government affairs, saying that "almost every Federation committee has been affected by the high level of federal activity in water pollution control. The influence of government continues and we are still learning how to respond. Each committee tries to deal with the problem by looking for avenues of communication for their viewpoint. It is, at times, frustrating, and yet it is important that we continue the process."

Energy is a subject that has received its share of attention in recent years. President Sam Warrington expounded on the subject in his 1975 message. Sam pointed out that ours is an energy-consuming industry and that we should be concerned about use of energy in solving environmental problems. There is a relation between energy, the environment, and economic problems, and we recognize the interdependence of the three. We should not surrender our principles, but rather should promote a responsible approach to environmental matters and economic and energy issues.

Treatment works operation was a favorite subject of nearly all past presidents. Harris Seidel, in 1964, titled his President's Message "The Operator—Today's Unforgotten Man" and said, "The outlook for operators in this vital public service field of water pollution control is stimulating. For those who are able to learn and willing to work, the future is bright indeed." Those remarks were equally appropriate in 1976. On this same subject, Robert Shaw, in 1966, pointed out that a successful water pollution control program depends on well-trained, responsible, devoted, and well paid treatment plant operators. And Joseph Hanlon, in 1970, pointed with pride to an ambitious program called MANFORCE (MANpower FOR a Clean Environment).

The name disappeared, but the concept continued to bear fruit with the Federation's expanding educational program and *Deeds & Data* publication. The relevance of this operator-oriented publication depends on the material printed in it. It provides an opportunity for operator members to get involved and to be of service to other operators.

Over the years the Federation and its Personal Advancement Committee continued to search for new tools to train operators, with considerable success. The Technical Practice Committee, too, was involved in the Federation's training and educational program. All of the Federation Manuals of Practice, the product of the Committee's activities, were useful in WPCF education and training activities.

Over the 15 years preceding my term of office, every president acknowledged the importance of the operator without differentiating between persons working in a treatment plant and those working on collection systems. This was an over-simplification of the role of the collection system specialist. Although none of my predecessors wrote specifically about collection system operators, their actions resulted in increased attention to collection systems.

In 1971, an *ad hoc* Committee on Wastewater Collection Systems was established and in 1972 became a standing committee. A collection system award and the qualifications for it were established in October 1973. Thus the collection system practitioner, after many years of leadership in the field, began to be recognized as a separate and distinct entity from those who concern themselves with treatment. Safety, monitoring, sewer-use regulation, overflows, storm water contamination, and a maze of new collection system technologies have emerged.

Laboratory scientists and technicians, too, received more and more recognition as an entity separate from those involved with treatment plant operation. Certification of these individuals was still in its infancy, but the push was to secure recognition of these specialized individuals. WPCF's participation in the publication of "Standard Methods" has exemplified the Federation's interest in laboratory technology. Comments on that participation appear in the discussion of Federation publications in Chapter VI. Much of the credit for WPCF's ultimate participation as a co-publisher of "Standard Methods" goes to Dr. W.D. (Hap) Hatfield, Federation president in 1959-60, who wrote much of the material on sewage analysis that appeared in the 9th edition in 1946 before WPCF was a partner in the publishing venture, and to Hatfield's protege, George Symons, who edited the 9th edition for APHA and AWWA.

Management of wastewater facilities was a subject with which Paul Haney challenged the Federation in 1969. Paul advocated that the Federation and its Member Associations should become deeply concerned about the role of management of wastewater utilities and several problem areas to be dealt with: Is the poor safety record in treatment plants a result of employee failures or failure of management? Is management ready to deal with employee relation problems? The ever-increasing cost of constructing, operating, and maintaining facilities require increased fiscal management and WPCF needs to improve its service to managers.

Publications were an important concern to all past presidents as they served the Federation. In 1965, Al Steffen pointed out the tremendous amount of new information generated in the water pollution control industry, to wit: the total amount of information available doubles every eight years. Steffen challenged the Federation to come up with ways to search this information pool for relevant material and to disseminate it to the membership.

The dissemination of information has taken many forms over the years. New sessions were added, almost every year, to annual conference programs. Additional pages in the journal were a yearly occurrence. The annual Literature Review, prepared by the Research Committee, has almost exceeded the convenience of publishing in a single issue.

Student Activities was not referred to in preceding presidential messages. The Federation, of late, has moved into this area. Two highly successful student-employer luncheons at the 1974 and 1975 Conferences pointed out the need to establish this affair as an annual feature of the Conferences.

Other activities which I recall as being important are the new Metrication Committee; the re-established Disinfection Committee, the Committees on Federation Organization, Marine Water Quality, Water Reuse, Public Relations, and Human Resources, the International Committee, and the Industrial Waste Committee.

It was Sid Berkowitz, who in his 1968 message said: "Let us then, using our past history, together with knowledge of the present and trained professional judgment, project our future."

When asked "what do you recall about your membership and your service as a Federation officer," I had no hesitancy in saying, "Besides the activities and ideas of other presidents that I observed before I became an officer, I remember this: The past year was the busiest and most demanding year of my life." It was also the most rewarding, for which I thank the members of the Federation, my fellow officers, the members of the Executive Committee, the Board of Control, the other nearly 1,000 volunteers that served on Committees of the Federation, and the thousands that serve our member associations. I have been impressed over the years with the quality of people on the Federation staff. They have provided outstanding service and have demonstrated an unusual amount of the flexibility so necessary in operating an organization such as the Federation.

Paul Haney, who first proposed the preparation of a Federation history, appropriately titled his 1969 message, "End and Beginning," signifying that the Federation is a continuing entity. Looking to Horace Smith, President of the WPCF during its 50th year, and Richard Engelbrecht and Martin Lang, who will be the first two presidents for the second half century, makes it easy to express the thoughts of all past presidents that "the best is yet to come."

Indianapolis, Ind.
November 1976

Horace L. Smith

Federation President 1976-77

As president of the Water Pollution Control Federation during its 50th Anniversary year, I was given the opportunity of providing an input to the documentation of its history. With a choice between contributing a foreword or epilog to the chapter on reminiscences, I chose the latter inasmuch as I shall be the president of record at the time the Federation history is published.

When I became president in Minneapolis on October 7, 1976, my remarks to the Board of Control were based on my analysis of Vic Wagner's Presidential Message, published in the September 1976 Annual Conference issue of the *Journal*. Vic had summarized and elaborated on the highlights of the previously published Presidential Messages. His effort disclosed an outline for the continuing development and progress of Federation programs and projects. When itemized, it is a list of basic needs identified by the Past Presidents of the Federation. These identifiable needs are:

- Acceleration of membership growth.
- An organizational structure to stimulate the interest and participation of a diversified membership.
- Increased activity in Member Associations.
- Improved training and expanded programs for wastewater treatment.
- Broader scope for Federation programs.
- Improved dissemination of technical information to the field and continued education of the public.
- Increased student activities and involvement.
- Recognition of the interdependence of the energy, economic, and environmental factors.
- Involvement in governmental affairs and learning how to respond to its influence.
- Acquiring responsiveness to special concerns.

Speculation on current problems and concerns with respect to water pollution control led me to listing some additional needs, including:

- Expanding coordinative and cooperative roles with other professional and technical organizations such as AWWA, APWA, and ABC.
- Ensuring the organizational continuity of dynamic programs and projects.
- Continuing reassessment, redefinition, and restatement of organizational goals, objectives, and policies.
- Correlation of the responsibilities and activities of the various Federation committees and providing direction for the accomplishment of their work programs and projects.

In addition to these general programs, there are several special concerns in which the Federation should be interested:

- Control of toxic substances.
- Management of residual wastes.

- Control of nonpoint source pollution.
- Water quality standards and methods and techniques of surveillance and monitoring.
- Planning for water quality management.
- Development of operational adequacy of "on line" wastewater treatment facilities.

Another issue to which the Federation must address itself is that of limited representation. Its programs have been concentrated in the municipal waste sector. Program efforts need to be expanded in the industrial waste sector. New programs must be developed for the agriculture, silviculture, and marine sectors of waste management. Only if these efforts are made will the Federation remain as the organization that represents the whole spectrum of water pollution control.

This composite itemization of concerns and needs represents the factors or components of the WPCF equation for the treatment of wastewater and control of water pollution.

The escalation of pollution abatement legislation and regulations and public and governmental clamor for accelerated advancement of the state of the art of water pollution control has made it most difficult to keep the components in perspective and balance. To accomplish this goal will be the Federation's challenge and responsibility for the future. The degree of change must be stabilized through the implementation of effective, efficient, and economical water pollution control programs. Projects must be harmonized and balanced with respect to resources, technology, legislation, and regulation.

As an epilog to these various reminiscences, may I repeat Shakespeare's words: "the past is but prologue to the future." The Federation must continue to attract to its membership the expertise necessary to meet its challenges and to discharge its responsibilities as stated in its pledge to provide leadership and guidance to *all constructive efforts* that contribute to the control of water pollution.

Indianapolis, Ind.
December 1976

Appendix

The following items are included in this Appendix to the Federation's history because they have been referred to elsewhere, because they are of special interest, or because they recreate a sense of what the early Federation was like.

- List of Presidents, terms of office, conference cities, dates, and registration
- Members of Committee of One Hundred (1927)
- Members of Implementing Committee (1928)
- Advertising Contract for *Sewage Works Journal* (1928)
- Table of Contents for "Modern Sewage Disposal" (1938)
- Letter from Earl Waterman to L.J. Murphy re. Federation Meeting, Chicago, Ill. (1940)
- Letter from C.A. Emerson to Association Secretaries (1941)
- Auxiliary Organizations (starting 1941)
- *FSWA Convention News*, Vol. 3, Cleveland Conference (1942)
- The *Federation Daily*, Vol. 4, Chicago Conference (1943)
- WPCF Awards and Medals (starting 1943)
- *Sewage Works Journal*, editorial, "Thanks to the Chemical Foundation" (March, 1944)
- Special Report on FSWA Research Activities (1953)
- "This is Your Life" script for H.E. (Ted) Moses (1955)
- Letter from H.E. Moses to Dr. Mattison re 1955 FSWA Conference
- Statement of Policy
- Letter from Walter A. Sperry (1969)
- Abbreviated History of Selected Committees of WPCF
- Water Reuse, Joint Resolution by WPCF and AWWA
- WPCF Growth Trends

List of WPCF Presidents and Annual Conferences

Annual Conference

President	Term of Office	City	Dates	Registration Total	Ladies
Charles A. Emerson*	1928-41	Chicago, Ill.	Oct. 3-5, 1940	556	
		New York, N.Y.	Oct. 9-11, 1941	559	
Arthur S. Bedell*	1941-42	Cleveland, Ohio	Oct. 22-24, 1942	412	26
George J. Schroepfer	1942-43	Chicago, Ill.	Oct. 21-23, 1943	612	63
A M Rawn*	1943-44	Pittsburgh, Pa.	Oct. 12-14, 1944	524	35
Albert E. Berry	1944-45	Chicago, Ill.	Oct. 17-18, 1945	Bd. Mtg. only	
John K. Hoskins*	1945-46	Toronto, Ont.	Oct. 7-9, 1946	812	202
Francis S. Friel*	1946-47	San Francisco, Calif.	July 21-24, 1947	964**	252
George S. Russell*	1947-48	Detroit, Mich.	Oct. 18-21, 1948	749	112
Victor M. Ehlers*	1948-49	Boston, Mass.	Oct. 17-20, 1949	821	142
Arthur H. Niles*	1949-50	Washington, D.C.	Oct. 9-12, 1950	904	151
Ralph E. Fuhrman	1950-51	St. Paul, Minn.	Oct. 8-11, 1951	718	124
Earnest Boyce	1951-52	New York, N.Y.	Oct. 6-9, 1952	1,152	209
E. Sherman Chase*	1952-53	Miami, Fla.	Oct. 13-16, 1953	805	199
Louis J. Fontenelli	1953-54	Cincinnati, Ohio	Oct. 11-14, 1954	1,103	205
David B. Lee*	1954-55	Atlantic City, N.J.	Oct. 10-13, 1955	1,345	294
George W. Martin*	1955-56	Los Angeles, Calif.	Oct. 8-11, 1956	1,030	242

* Deceased
** Joint meeting with AWWA; combined attendance 1,962

President	Term of Office	City	Annual Conference Dates	Registration Total	Ladies
Emil C. Jensen	1956-57	Boston, Mass.	Oct. 7-10, 1957	1,363	301
Kenneth S. Watson	1957-58	Detroit, Mich.	Oct. 6-9, 1958	1,349	238
William D. Hatfield	1958-59	Dallas, Tex.	Oct. 12-15, 1959	1,272	237
Mark D. Hollis	1959-60	Philadelphia, Pa.	Oct. 2-6, 1960	1,720	325
Ray E. Lawrence*	1960-61	Milwaukee, Wis.	Oct. 8-12, 1961	2,056	328
Harry E. Schlenz*	1961-62	Toronto, Ont.	Oct. 7-11, 1962	2,277	503
Jack E. McKee	1962-63	Seattle, Wash.	Oct. 6-10, 1963	1,951	363
Harris F. Seidel	1963-64	Bal Harbour, Fla.	Sept. 27-Oct. 1, 1964	2,054	464
A.J. Steffen	1964-65	Atlantic City, N.J.	Oct. 10-14, 1965	3,461	582
Robert S. Shaw	1965-66	Kansas City, Mo.	Sept. 25-30, 1966	3,483	443
Arthur D. Caster	1966-67	New York, N.Y.	Oct. 8-13, 1967	4,318	571
Sidney A. Berkowitz	1967-68	Chicago, Ill.	Sept. 22-27, 1968	4,806	624
Paul D. Haney	1968-69	Dallas, Tex.	Oct. 5-10, 1969	4,126	612
Joseph B. Hanlon	1969-70	Boston, Mass.	Oct. 4-9, 1970	5,037	714
Arthur F. Vondrick	1970-71	San Francisco, Calif.	Oct. 3-8, 1971	5,702	1,103
Joseph F. Lagnese, Jr.	1971-72	Atlanta, Ga.	Oct. 8-13, 1972	7,055	1,213
J. Floyd Byrd	1972-73	Cleveland, Ohio	Sept. 30-Oct. 5, 1973	6,261	707
John D. Parkhurst	1973-74	Denver, Colo.	Oct. 6-11, 1974	7,913	1,432
Sam Warrington	1974-75	Miami Beach, Fla.	Oct. 5-10, 1975	8,098	1,321
Victor G. Wagner	1975-76	Minneapolis, Minn.	Oct. 3-8, 1976	8,716	1,116
Horace L. Smith	1976-77	Philadelphia, Pa.	Oct. 2-7, 1977	—	—

Federation of Sewage Works Associations
Committee of One Hundred—Appointed July 15, 1927

C.A. Emerson, Jr., Executive Chairman

Sub-Committee on Organization of Federation

Langdon Pearse (Chairman), Chicago, Ill.
C.M. Baker, Madison, Wis.
Earnest Boyce, Lawrence, Kan.
J.W. Bugbee, Providence, R.I.
L.M. Clarkson, Atlanta, Ga.
W.R. Copeland, Hartford, Conn.
H.P. Croft, Trenton, N.J.
W.W. DeBerard, Chicago, Ill.
W.H. Dittoe, Youngstown, Ohio
Frank C. Dugan, Louisville, Ky.
V.M. Ehlers, Austin, Tex.
Jos. W. Ellms, Cleveland, Ohio
Harry F. Ferguson, Springfield, Ill.
E.L. Filby, Jacksonville, Fla.
L.S. Finch, Indianapolis, Ind.
Stephen DeM. Gage, Providence, R.I.
George G. Gascoigne, Cleveland, Ohio
C.G. Gillespie, Berkeley, Calif.
S.A. Greeley, Chicago, Ill.
G.H. Hazelhurst, Montgomery, Ala.
C.A. Holmquist, Albany, N.Y.
H.B. Hommon, San Francisco, Calif.
Theodore Horton, Albany, N.Y.
H.E. Jordan, Indianapolis, Ind.
Richard Messer, Richmond, Va.
R.B. Morse, Hyattsville, Md.
John H. O'Neill, New Orleans, La.
Malcolm Pirnie, New York, N.Y.
Edward D. Rich, Lansing, Mich.
Ellis S. Tisdale, Charleston, W.Va.
E.D. Walker, State College, Pa.
H.A. Whittaker, Minneapolis, Minn.
Chester G. Wigley, Maplewood, N.J.
H.C. Woodfall, Atlanta, Ga.
Edward Wright, Boston, Mass.

Sub-Committee on Finance Program

George W. Fuller (Chairman), New York, N.Y.
Frank Bachmann, Chicago, Ill.
H.N. Calver, New York, N.Y.
Louis J. Dublin, New York, N.Y.
A.D. Flinn, New York, N.Y.
Charles G. Hyde, Berkeley, Calif.
M.W. Loving, Chicago, Ill.
S.F. Miller, New York, N.Y.
David Morey, Jr., Dallas, Tex.
W.J. Orchard, Nutley, N.J.
George H. Shaw, Philadelphia, Pa.
M.B. Tark, Philadelphia, Pa.
Thos. F. Wolfe, Chicago, Ill.

Sub-Committee on Prospective Research Papers

H.W. Streeter (Chairman), Cincinnati, Ohio
H.G. Baity, Chapel Hill, N.C.
A.M. Buswell, Urbana, Ill.
H.W. Clarke, Boston, Mass.
F.E. Daniels, Harrisburg, Pa.
L.H. Enslow, New York, N.Y.
A.L. Fales, Boston, Mass.
Gordon M. Fair, Cambridge, Mass.
Max Levine, Ames, Iowa
Isador W. Mendelsohn, Chicago, Ill.
F.W. Mohlman, Chicago, Ill.
E.B. Phelps, Ridgewood, N.J.
W. Rudolfs, New Brunswick, N.J.
Abel Wolman, Baltimore, Md.

Sub-Committee on Prospective Operative Data

C.E. Keefer (Chairman), Baltimore, Md.
Kenneth Allen, New York, N.Y.
M.N Baker, New York, N.Y.
C.K. Calvert, Indianapolis, Ind.
G.L. Fugate, Houston, Tex.
John H. Gregory, Baltimore, Md.
Henry N. Heisig, Milwaukee, Wis.
Glenn D. Holmes, Syracuse, N.Y.
Clarence B. Hoover, Columbus, Ohio
J.K. Hoskins, Cincinnati, Ohio
W.T. Knowlton, Los Angeles, Calif.
W.S. Mahlie, Fort Worth, Tex.
John F. Skinner, Rochester, N.Y.

Sub-Committee on Coordination of Local Groups

H.P. Eddy (Chairman), Boston, Mass.
F.A. Barbour, Boston, Mass.
Edward Bartow, Iowa City, Iowa
H.W. Dodds, New York, N.Y.
Wellington Donaldson, New York, N.Y.
John R. Downes, Bound Brook, N.J.
E.G. Eggert, Austin, Tex.
C.A. Emerson, Philadelphia, Pa.
Arthur E. Gorman, Chicago, Ill.
George T. Hammond, Brooklyn, N.Y.
Paul Hansen, Chicago, Ill.
W.D. Hatfield, Decatur, Ill.
J.J. Hinman, Iowa City, Iowa
T. Chalkley Hatton, Milwaukee, Wis.
W.W. Horner, St. Louis, Mo.
Theo. Lafreniere, Montreal, Canada
J. Horace McFarland, Harrisburg, Pa.
A.P. Miller, Washington, D.C.
H.E. Miller, Raleigh, N.C.
Arthur Ringland, Washington, D.C.
Warren J. Scott, Hartford, Conn.
Lent D. Upson, Detroit, Mich.
F.H. Waring, Columbus, Ohio
R.S. Weston, Boston, Mass.
C.-E.A. Winslow, New Haven, Conn.

Federation of Sewage Works Associations Implementing Committee, Appointed March 8, 1928

C.A. Emerson, Jr., Executive Chairman

Organization Committee

H.W. Streeter (Chairman), Cincinnati, Ohio
Robert Cramer, Milwaukee, Wis.
C.H. Currie, Webster City, Iowa
F.H. Dryden, Salisbury, Md.
V.M. Ehlers, Austin, Tex.
W. Scott Johnson, Jefferson City, Mo.
C.D. Maguire, Columbus, Ohio
Raymond O'Donnell, State College, Pa.
W. Rudolfs, New Brunswick, N.J.

Finance Committee

W.J. Orchard (Chairman), Newark, N.J.
H.P. Croft, Trenton, N.J.
L.J. Murphy, Ames, Iowa
G.H. Radebaugh, Urbana, Ill.
L.F. Warrick, Madison, Wis.

Publication Committee

John R. Downes (Chairman), Bound Brook, N. J.
A.L. Fales, Boston, Mass.
C.G. Gillespie, Oakland, Calif.
C.A. Holmquist, Albany, N.Y.
Max Levine, Ames, Iowa
A.C. Magill, Cape Girardeau, Mo.
W.S. Mahlie, Fort Worth, Tex.
Richard Messer, Richmond, Va.
A.P. Miller, Chevy Chase, Md.
F.W. Mohlman, Chicago, Ill.
Langdon Pearse, Chicago, Ill.
Warren J. Scott, Hartford, Conn.
E.D. Walker, State College, Pa.
F.H. Waring, Columbus, Ohio
Abel Wolman, Baltimore, Md.

Coordination Committee

Kenneth Allen (Chairman), New York, N.Y.
P.N. Daniels, Trenton, N.J.
L.S. Finch, Indianapolis, Ind.
T. Chalkley Hatton, Milwaukee, Wis.
J.J. Hinman, Iowa City, Iowa
John F. Skinner, Rochester, N.Y.
W.L. Stevenson, Harrisburg, Pa.
A.H. Wieters, Des Moines, Iowa

Advertising Contract

SEWAGE WORKS JOURNAL

Business Manager's Office
85 Beaver Street,
NEW YORK, N.Y.

You are hereby authorized to insert our advertisement in the Sewage Works Journal, beginning with ....
(month)
issue, to occupy space of ....
(full page) (half page) (etc.)
page quarterly for a period of one year and thereafter until otherwise ordered in writing; for which we agree to pay on publication at the rate of $.................... per insertion.

For good and sufficient reason this agreement may be terminated by the advertiser by giving notice in writing thirty days in advance of date of issue, and on payment of the short time rates for the insertions made, as shown by the rates on this form. When change of copy is not provided before your closing date the preceding advertisement is to be repeated.

Signed .... (Company)

By ....

Address ....

....

NET ADVERTISING RATES
SEWAGE WORKS JOURNAL
85 Beaver Street, New York City

Contract basis of 4 quarterly insertions:	Single Issues:
1 page .... $300.00 a year	1 page ... $100.00 each
½ page .... 175.00 "	½ page ... 65.00 "
¼ page .... 100.00 "	

PUBLISHED QUARTERLY—JANUARY, APRIL, JULY, OCTOBER
First forms close 1st of preceding month, last forms 15th of preceding month. Type page size 4½ x 7½ inches.

Contents of "Modern Sewage Disposal"
Tenth Anniversary Book of FSWA

CONTENTS

Foreword

	C.A. Emerson, Jr.	The Federation and Its Relationship to Sewage Disposal	vii

Introduction

I.	C.G. Hyde	A Review of Progress in Sewage Treatment During the Past Fifty Years in the United State	1

Sewage Treatment Practice

II.	Langdon Pearse	Functional Outline of Processes of Sewage Treatment	16
III.	Samuel A. Greeley	Sedimentation and Digestion in the United States	28
IV.	Karl Imhoff	Sedimentation and Digestion in Germany	47
V.	W.E. Stanley	The Trickling Filter in Sewage Treatment	51
VI.	F.W. Mohlman	Twenty-five Years of Activated Sludge	68
VII.	W. Donaldson	Chemical Treatment of Sewage	85
VIII.	Linn H. Enslow	Chlorine in Sewage Treatment Practice	98
IX.	George B. Gascoigne	Mechanical Equipment in Modern Sewage Treatment	110
X.	Morris M. Cohn	Utilization and Disposal of Sewage Sludge	115
XI.	W.D. Hatfield	The Works Laboratory	126
XII.	C.A. Holmquist and C.C. Agar	Bettering Sewage Plant Operation	132
XIII.	Max Levine	Bacteriological Aspects of Sewage Treatment	139
XIV.	Earle B. Phelps	Bio-Chemistry in Sewage Treatment	158
XV.	W.C. Purdy	Limnological Aspects of Sewage Disposal	171
XVI.	N.T. Veatch, Jr.	The Use of Sewage Effluents in Agriculture	180

XVII.	H.W. Streeter	Disposal of Sewage in Inland Waterways	191
XVIII.	A.K. Warren and A M Rawn	Disposal of Sewage into the Pacific Ocean	202
XIX.	A.D. Weston	Disposal of Sewage into the Atlantic Ocean	209

Sewage Research

XX.	Willem Rudolfs	Sewage Research	219
XXI.	Harold E. Babbitt	Research in Sewage Treatment at Educational Institutions in the United States	237
XXII.	C.E. Keefer	Administration of Sewage Treatment Works	248
XXIII.	Gordon M. Fair	Mathematical Aspects of Sewage Research	259

Regional and National Aspects

XXIV.	C.-E.A. Winslow	Pioneers of Sewage Disposal in New England	276
XXV.	Abel Wolman	State and Other Governmental Function in the Control and Abatement of Water Pollution in the United States	285
XXVI.	H.T. Calvert	Governmental and Local Functions in the Control of Sewage Disposal in Great Britain	298
XXVII.	J.H. Garner	Sewage Treatment in England	302
XXVIII.	E.J. Hamlin and H. Wilson	The Sewage Disposal Problem in South Africa	315
XXIX.	H.J.N.H. Kessener	Sewage Treatment in the Netherlands	325
XXX.	A. Heilmann	Sewage Treatment in Germany	333

Industrial Wastes

| XXXI. | L.F. Warrick | The Prevalence of the Industrial Waste Problem | 340 |
| XXXII. | Robert Spurr Weston | Treatment and Disposal of Industrial Wastes | 351 |

THE STATE UNIVERSITY OF IOWA
College of Engineering
IOWA CITY, IOWA

October 8, 1940

Professor Lindon J. Murphy
Secretary-Treasurer
Iowa Wastes Disposal Association
Ames
Iowa

Dear Pat:

I thank you for your letter of October 2 advising me that the Executive Committee had voted to pay $15.00 toward the expenses of the Directors of the Iowa Wastes Disposal Association for attendance at a meeting of the Board of Control, Federation of Sewage Works Associations held in Chicago, Saturday, October 5. Both Dr. Levine and I attended the meeting.

I did not receive a copy of the report of the Committee on Reorganization and Extension prior to going to Chicago. Dr. Levine was familiar with it. There was a good deal of discussion on controversial points in the proposed constitution and by-laws but these were fairly well ironed out but not entirely so. There were a number of members of the original committee who were going on to Detroit for the meeting of the A.P.H.A. this week. They were appointed a sub-committee to revise the original report. This was to be done early this week in Detroit. After revision, copies of the proposed constitution and by-laws are to be sent to each of the member associations and action of some sort must be obtained before November 15. The reason for this is that 60 day's notice is to be given on the proposed changes in the constitution. The Annual Meeting of the Board of Control at which it is hoped that final action can be taken, will be held in New York about January 15.

If the Annual Meeting of the Iowa Wastes Disposal Association is to be held before the middle of November the whole plan can be brought up for discussion and action. If the meeting is to be later it is probable that action will have to be taken by our Executive Committee. It appears to me that this matter is being hurried but I am told that for financial reasons it is necessary to get the new set-up into operation as soon as it can reasonably be accomplished.

With kind personal regards, I am

Yours very truly,
Earle L. Waterman

FEDERATION OF SEWAGE WORKS ASSOCIATIONS

Office of the President
233 Broadway
New York City

March 8, 1941

TO THE SECRETARIES OF THE MEMBER ASSOCIATIONS:

Gentlemen:

I take pleasure in advising you that the central office of the Federation has been opened and is now functioning under the direction of W.H. Wisely, Executive Secretary, whose appointment was authorized by the Board of Control at the annual meeting on January 15. His address is P.O. Box 18, Urbana, Illinois.

Mr. W.W. DeBerard of Chicago has qualified as Treasurer and taken over the management and custody of the funds of the Federation.

Mr. H.E. Moses of Harrisburg, who has served so faithfully as Secretary-Treasurer since the inception of the Federation in 1928, will continue temporarily as Honorary Secretary and give us the benefit of his advice and assistance.

Doctor F.W. Mohlman of Chicago will continue as Editor and has definite plans for expanding the Operators' Section of the JOURNAL.

Mr. Arthur A. Clay of Chemical Foundation, our former Business Manager, becomes Advertising Manager and will continue in charge of the advertising and the mechanical production of the JOURNAL for the remainder of this year. Other duties formerly performed by Chemical Foundation have been divided between Messrs. Wisely and DeBerard.

The Federation has been incorporated under the Illinois law regulating "Corporations not for Pecuniary Profit."

In conformity with these changes, will you please address all correspondence relating to membership, JOURNAL subscriptions, dues, and other general Federation matters to Mr. Wisely, instead of to Mr. Moses as formerly. Correspondence relating directly to articles in the JOURNAL should, of course, continue to be sent to Doctor Mohlman in Chicago.

Mr. Wisely hopes to be able to attend a number of the member association meetings during the coming months, and I only wish our finances were such that he could make all of you a visit, but it is not possible this first year. Just as soon as the date and place of your 1941 meeting or meetings have been determined, please send the information to Mr. Wisely so that our central office files will be complete and up-to-date.

A good many of the member associations will find it necessary to amend their Constitutions and By-Laws to conform to the new Constitution and By-Laws of the Federation. Correspondence relating to such changes, together with the text of the same, should go to Mr. H.W. Streeter,

Chairman of our Organization Committee, whose address is care of the U.S. Public Health Service, Third and Kilgour Streets, Cincinnati, Ohio.

Since our Board of Control meeting in January numerous letters have been received congratulating the Federation on the culmination of the long considered plans for expansion of activities, and I am pleased to report that everything appears to be moving along nicely toward fulfillment. There is still much to be accomplished, but with full cooperation the Federation can continue to grow and be of greater service than ever before. I am counting on all of you for loyal support in the honest belief that every member of every member association—wherever he is located and however he is situated—will be benefited.

                Very truly yours,
                FEDERATION OF SEWAGE WORKS ASSOCIATIONS
                C.A. Emerson
                President

Auxiliary Organizations

The Water Pollution Control Federation has one official auxiliary and one loosely-associated auxiliary sponsored by a number of individual associations. These are The Quarter Century Operators' Club and the Select Society of Sanitary Sludge Shovelers. There is no ladies auxiliary, although there are a number of lady members. Ladies' activities for the wives have been a very important part of the Federation's annual conferences as well as the Member Association meetings. The Iowa Association does have a Ladies Auxiliary.

THE QUARTER CENTURY OPERATORS' CLUB

This organization was conceived and started in 1941 by Frank Woodbury Jones. In 1947-48 it was formally established with the approval of the Federation Board. Jones was then a member of the consulting firm of Havens and Emerson in Cleveland, Ohio, but he had been superintendent of the sewage treatment plant in Ohio prior to 1916 which, of course, was 12 years before the Federation was founded.

Frank Jones proposed that the Club be an informal group of active (later also corporate) members of any member association of the Federation; he also proposed that the criteria for membership be full-time resident charge of wastewater treatment plant operation for at least one year, on a full-time resident basis, twenty-five years prior to the date of admission to the Club.

This "full-time resident charge" criteria was the source of some good-natured beefing by chemists, engineers, and shift-operators, who claimed that the organization should really be called the "Quarter Century Superintendents' Club" not "Operators' Club."

The Club meets, and its members are recognized at the Federation Luncheon during each annual conference. Persons who become eligible have their names added to the roster by a Registrar. The first Registrar was Frank Woodbury Jones. He was followed by Morris M. Cohn, long-time chairman of the Federation's Sewage Works Practice Committee. Henry Van der Vliet succeeded Dr. Cohn, and Art Caster succeeded Van der Vliet.

The first roster included eight names:
- Julius Bugbee, Providence, R.I.
- Stuart E. Coburn, Boston, Mass.
- John Downes, Plainfield, N.J.
- Charles C. Hommon, Canton, Ohio
- Frank Woodbury Jones, Cleveland, Ohio
- Roy Lanphear, Worcester, Mass.
- Paul Molitor, Sr., Madison-Chatham, N.J.
- William Piatt, Durham, N.C.

In the years after 1941, the number of Club members grew slowly at first, attesting to the fact that sewage treatment plant operation (as it was termed in the 1940's) had not been widespread in the late teens and early twenties. By the 1970's many of the early members of the Club had retired or passed away. The following data give an indication of the growth not only of the Quarter Century Club but of wastewater treatment plants in this country.

QUARTER CENTURY OPERATORS CLUB

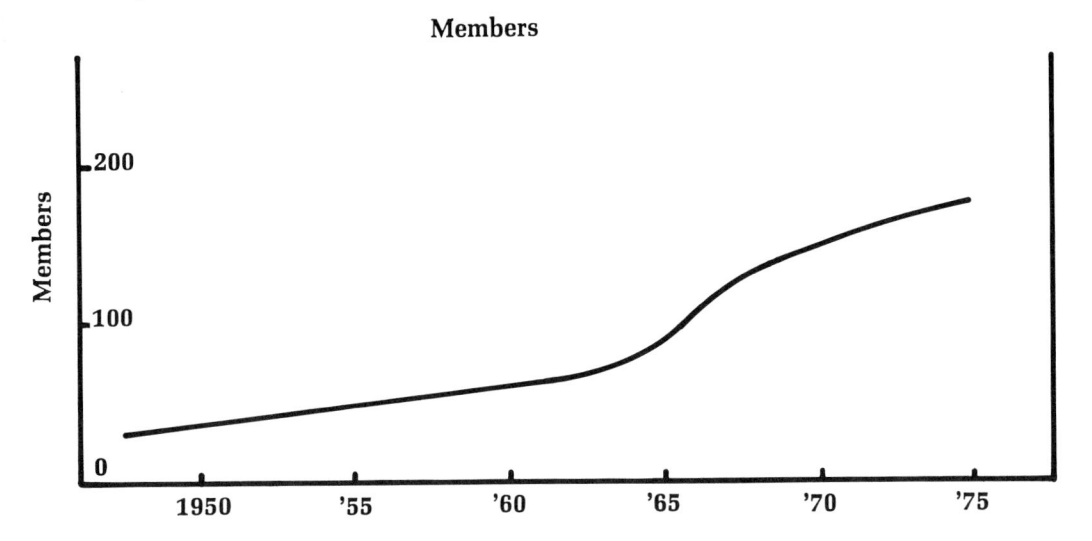

Membership is recognized by the Federation each year in the program of each annual conference which contains a printed roster of members.

THE SELECT SOCIETY OF SANITARY SLUDGE SHOVELERS

According to information provided by Art Vondrick (Federation President 1970-71), there are nine chapters of this Select Society. They are loosely associated; each exists only within the jurisdiction of a member association of the Federation; each inducts a limited number of members at some time during the annual meeting of its parent association; each provides new members with a gold/silver tie bar in the shape of a round-nose shovel or a spade; and each is a fun group.

Different chapters use different initiation ceremonies; there is no formal written outline or set procedure. Some chapters make the ceremony a little more serious than others and eliminate the frivolity; others do it all in fun.

All the chapters except Pennsylvania, Texas, Florida, and Central States use identical gold shovels as their emblems. California engraves the year in the bowl of the shovel; Rocky Mountain engraves a large "R". The Texas shovel is smaller and has a 5-pointed star (the Lone Star) in the bowl with an "S" opposite each point of the star. The Central States shovel is larger than the rest; the Pennsylvania and Florida shovels are spade-like, but are silver or pot metal with different symbols embossed on them.

There has been considerable disagreement as to which group was created first. Pennsylvania started the High Hat Society in 1937; it used the words "Sludge Shovelers Society" in its initiation ceremony and later became known as the Ted Moses Sludge Shovelers Society.

The Arizona Association documents its history from October, 1940, when the idea was conceived by A.W. (Dusty) Miller* and F. Carlyle Roberts, Jr. "to recognize the fact that many members do not receive the coveted Bedell Award, the Fuller Award (of AWWA) or become an Arizona Association president, but nevertheless contribute in some outstanding measure."

According to information available at the time of compilation of this Appendix, the list of Select Societies of Sanitary Sludge Shovelers consisted of:

        Pennsylvania WPCA (1937)
        Arizona Wtr & Sew. Works Assn. (1940)
        Texas WPCA (1954)
        California (1955)
        Florida WPCA (1956)
        Rocky Mountain WPCA (1963-1972)
        Utah WPCA (1970)
        Central States WPCA (1971)
        Kansas WPCA (1972)
        Missouri WPCA (1973)

---

* A.W. Miller and his wife died in an airplane crash in the Canary Islands on March 27, 1977.

The Federation does not have information on all of these groups but some items about the older ones are of interest, as much perhaps for the men who started the organization as for the details of the group's operations.

The Ted Moses Sludge Shovelers Society

The budget for Pennsylvania Sewage Works Assn. meeting at State College, Pa. in Aug. 1937 was short of funds for paid entertainment at the banquet. As a substitute, L.D. Matter and Bernie Bush, both of Pennsylvania Health Dept., shopped the theater costume area of Philadelphia and found a high hat and leather carrying case. H.E. (Ted) Moses made the presentation at the banquet. The exact text of his presentation is not recorded, but the explanation still given at each banquet is a good approximation. It reads as follows:

"A word of explanation is due some of our guests here tonight as to just what the High Hat ceremony means.

"In 1937 a new method of recognition for outstanding service to the sewerage profession was devised in the Pennsylvania Sewage Works Association, whereby recipients of an award were nominated by the Executive Committee, but were actually elected by those present at the annual dinner—after the candidates had successfully passed the exacting tests which the award calls for—in the presence of the assemblage.

"The High Hat Award began as entertainment, but is no longer so; and the recipient of the award today can be assured that his performance in the waste control field has been recognized as being outstanding by his contemporaries.

"The rules are very simple but exact, and are well known to most of you. After the candidate is identified, he is called to the platform, and in your presence required to say the magic words, "Sludge Shovelers Society," three times, with variations: in a normal tone of voice; a whisper; yell it out.

"You, as the Judges, are requested to note any imperfections of pronunciations or clarity of diction as the candidate performs, and signify your decision as to his having met the test at the proper time."

After each candidate had recited the words "Sludge Shoveler's Society," Ted Moses always said, "I now crown you with the High-Hat." Moses made the presentations from 1937 through 1956; he was followed by Bernie Bush (1957-58), Deac Matter (1959-65), and John Yenchko after 1966.

Each man who is crowned with the High-Hat has his name inscribed on the Hat, which is kept in its case by the Association's Secretary. In recent years, each recipient has received a silver tie-bar spade with the words "Ted Moses Sludge Shovelers Society" on it, the award having been so named after Mr. Moses' death.

The list of members of this group includes some greats, near greats, important personages, and just good Joes. All Federation presidents since 1937 have been crowned with the High-Hat in Pennsylvania.

Arizona Chapter, Select Society of Sanitary Sludge Shovelers

The following material, prepared by Art Vondrick in May 1971, is a combination history, bylaws, initiation ceremony, and certificate. This "document" is updated from time to time as needed:

"The Select Society of Sanitary Sludge Shovelers was originated to encourage what is now known as 'getting involved.' The measure of success of this organization is testified to by the fact that as of now there are members all over the United States, and some even in Canada, all of whom have been inducted, initiated, and integrated into the Society. You cannot join—you must be selected—on the basis of merit. This is a society for those who contribute their efforts, their time, and their energies.

"The rules are simple. Your efforts toward making your local Association a better one and making a better Annual Meeting come in the form of:

(1) Presenting at least two technical papers on the technical program.

(2) Presenting at least two nontechnical papers on the nontechnical program.

(3) Chairmanship of important productive committees.

(4) Other activities which in the opinion of the judges are worth recognition.

"The opinions of the judges are final. In case of ties, duplicate awards are not made, but the judges can be influenced.

"A cardinal Rule is that initiation, induction, and integration ceremonies must be conducted by a Sludge Shoveler.

"Tradition provides that the four senior members of the Society present at any meeting of the Association shall select three new chapter members who qualify by their activities at that meeting.

"Selection to membership in a chapter is in recognition of 'outstanding, meritorious service above and beyond the call of duty' to the Arizona Water & Pollution Control Association. Selection bestows the accolade of elevation 'on the official shovel to the highest ridge on the sludge bed, with the title of Select Sanitary Sludge Shoveler and all the honor, atmosphere, perquisites, and dignity appertaining thereunto.'

"There are no dues or officers except the Influent Integrator who is designated by the neutral pH7. He is elected by vote of the Select Sanitary Sludge Shovelers present at any meeting of the Association and serves until his successor is elected and installed. His duties are to record and report selections, present official certificates of elevation, bestow badges, and inform chapter members concerning the Society. A prerequisite of membership is the privilege of cleaning up before, during, and after meetings of the Association.

"Each certificate shall be signed by the Influent Integrator as pH7 and by any other twelve members for the remaining concentrations from pH1 to pH13. [*A copy of the certificate appears following this text.*]

"The badge is a shovel worn extending from the left breast pocket. The grip is made by curling the fingers as though around a shovel handle. The grand hailing sign is made by raising the grip head high, thumb on the left, and lowering smartly, thus symbolizing the close relation between the water and sewage in which the Association is interested.

"The signal of distress is a sweeping motion made with both hands as if shovelling. The chosen station of Select Sanitary Sludge Shovelers is at the opposite end of the meeting from the President of the Association. The pass word is derived from the first letters of the name and is pronounced, 'Sh-h-h.'

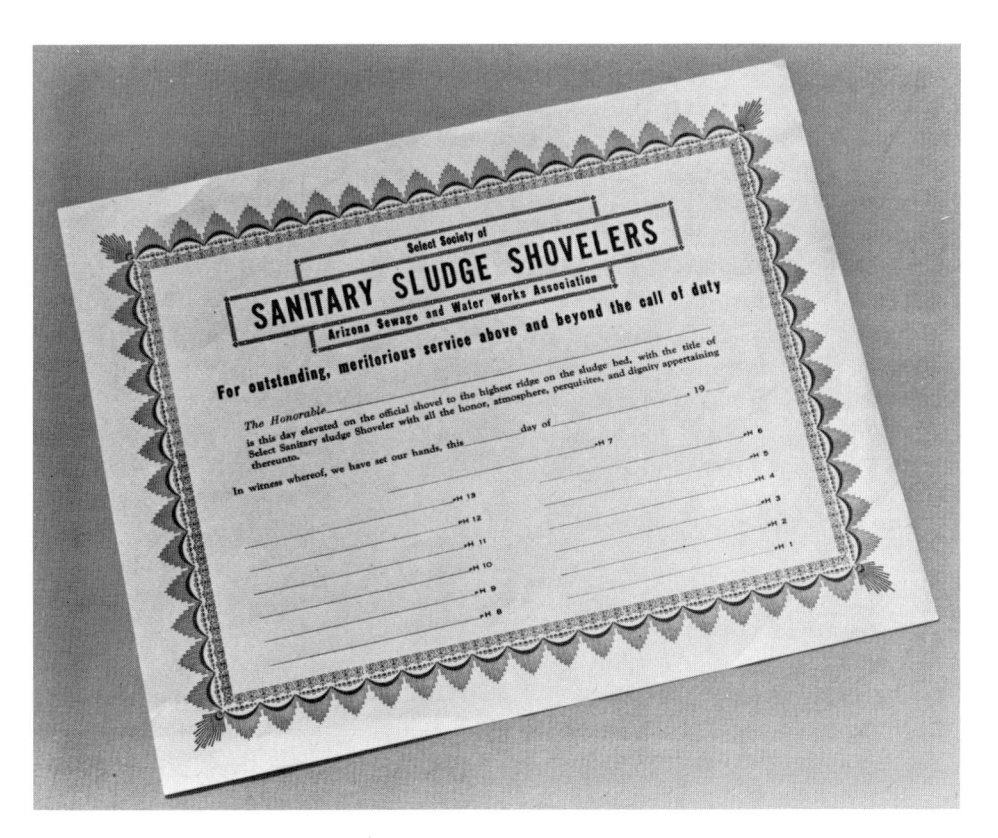

"The badge is only used at official initiation-integration ceremonies. The gold shovel emblem, worn either as a lapel pin, tie clasp, or other suitable item (as in the case of female members), should always be worn or displayed to indicate a member in good standing."

Texas Select Society of Sanitary Sludge Shovelers

According to the report on the 36th Texas Water Works Short School in March 1954, a committee set up an honorary organization within the Texas Sewage and Industrial Wastes Section:

"To be known as the Select Society of Sanitary Sludge Shovelers. Qualification for membership in the organization shall be the securing of at least five new members during the year.

"An attractive and appropriate emblem in the form of a gold shovel on a tie clasp was presented to the six persons who qualified for this distinction during the past year. Messrs. T.C. Ferris, Dean S. Matthews, and N.W. Black, all of whom were recipients of monetary prizes (in the 1954 membership contest) donated their prizes to assist in perpetuating the society."

Messrs. Roger Moehlman (Harris County Health Dept.) who was chairman of the membership contest committee for 1953-54 and A.C. Bryan of Houston had decided on the contest route to increase membership. Neither remembers which one "came up with the idea of the shovel" but they do recall that they decided to engrave the Texas Lone Star on the shovel and to put an "S" at each of three points of the star to represent "Sanitary Sludge Shovelers." The idea was dropped because it unbalanced the star, "so they decided to use the name Select Society of Sanitary Sludge Shovelers as this would place an "S" at each point of the star." This decision came 14 years after Arizona's organization and probably someone had heard the five "S" name before.

Rocky Mountain WPCA Society

The Rocky Mountain Association started presenting Sludge Shovel Awards in 1963 to qualified recipients. Until 1969, award eligibility was based solely on outstanding operations of facilities. From 1969 through 1971, the award was presented to individuals contributing to the success of annual meetings.

In September 1972, by Board action and approval of the membership, the Rocky Mountain Association's Select Society of Sludge Shovelers was officially created. With the first official ceremony, all past recipients of the Sludge Shoveler Award were included with two new members. The rules of eligibility adopted obviously were influenced by the Arizona Group.

Kansas Sludge Shovelers Society

In October 1973, John H. Bailey, Secy.-Treas. of the Kansas WPCA reported on the origin of its chapter as follows: "Kansas indeed has a Society of Sanitary Sludge Shovelers initiated in March of 1972. We are quite elated with the response the society has generated in our organization as it is truly a "fun" thing!

Bailey made this additional comment: "For the record, we had an extremely difficult time in obtaining the gold shovels for our presentations, and as a consequence, were loaned some from the Missouri Water Pollution Control Association which had obtained some through other channels. We now have enough sanitary sludge shovels to last for a long time. It is not too often that you find gold-plated shovels in jewelry stores."

FSWA CONVENTION NEWS*

Vol. 3, No. 1   Cleveland, Ohio   Thurs. October 22, 1942

The Third Annual Convention starts today. Register early, but not often. We wait with bated breath to see how many believed Wm. L. Havens' Convention Management Committee and Morris Cohn when they said, "It's a War Time Must."—"The National Conference on War Time Sanitation, in Cleveland."

— BEDELL, PRESIDENT —

And, lest you forget, this is also the Ohio Conference on Sewage Treatment.

— SCHROEPFER, VICE PRESIDENT —

We note on p. 1136 of the September issue of *Sewage Works Journal*, that Dr. Mohlman is trying to sell Homer (Pete to you) Wisely to the Federation—as if he needs to.

— EMERSON, PAST PRESIDENT —

And, speaking of Pete, you can't blame him this year for what appears here, for he delegated the column to Yours Truly, who once columned in a country weekly under the heading SYMÓN(S)IZING.

— DeBERARD, TREASURER —

"Oh, Easterly is Easterly, and Westerly is Westerly, and never the twain shall meet," wrote Kipling. On Saturday afternoon, we'll see how right he was.

— WISELY, EXECUTIVE SECRETARY —

Definition of a Symposium: (1) a drinking together, with conversation, and free interchange of ideas; (2) a conference at which a particular subject is discussed and opinions gathered. You'll find the first in several rooms tonight and the second tomorrow afternoon.

— MOHLMAN, EDITOR —

We note that the concert meister has become the conductor for one session and what an array of talent he has in his symphony—or didn't you know that "Cubic Centimeter" Larson was a fiddler?

— HAVENS, CONVENTION MANAGER —

And that other "Cubic Centimeter"—Agar, from New York—is in the "deep sooth" Sanitary Corpsing as are many others—we shall miss them all.

— DORR, GUEST SPEAKER —

We note that the boy with the double split name (none other than LeRoy Van Kleeck, himself) again turns the tables on the operators, on Saturday morning. He's following Clifton Fadiman's secret of success; he knows all the questions.

---

* Ed. Note: The "Convention News Dailies" were prepared late each night by George Symons, then Chief Chemist, Buffalo, N.Y., Sewer Authority, and were distributed at the session rooms and registration desk early each morning.

— ELLMS, OF CLEVELAND —

Highlight of the Inspection Trip Saturday at the Easterly Sewage Plant will be a buffet luncheon cooked in their own kitchen and served in their own dining room. A choice of three entrees and two desserts (Sludge Cake served at Southerly Plant) will be available to all comers. Transportation will be furnished. They already expect 75, says John J. Wirts, Superintendent. Best advice is to sign up when you register.

— RAWN, PROGRAMMER —

Don't know how many came in last night, but the N.Y. Central from the east carried the vanguard of the New York State Association, who are hoping for the man-miles cup this time. Remember last year? The meeting was held in their own state and they didn't touch the Central States record. Bet the Ohio Conference winds up the same way this time.

— BOY! LOOK AT THOSE EXHIBITS —

It's an old Indian proverb—"To make sure your smoke signals are read, mention the readers' names." If you don't see yours here today, look tomorrow.

— THIRTY —

FSWA CONVENTION NEWS

Vol. 3, No. 2   Cleveland, Ohio   Fri. October 23, 1942

NOTE CHANGE IN PROGRAM: To provide time for discussion of that all-important subject of priorities, the paper by J.W. Ellms, "Sewage Treatment Problems in Cleveland" was presented yesterday afternoon, and Rudolfs and Graff appear this morning at 9:00 A.M. This change leaves time at 9:45 A.M. for Pete Wisely to present Messrs. Neil, J.F. Van Steinberg, and Malcolm Slaght, of the War Production Board, Washington. They will discuss contemplated changes in priority procedures and answer questions from the floor.

— A.H. NILES, CHAIR. OHIO CONF. —

Familiar faces at the Pre-convention Get-together around The Beer-Keg: Milton Adams, Paul Hansen, Van Enloe, Profs. Kilcawley and Barnes, Marcus Tark, Don Bloodgood, Vance Crawford, Al Genter, "Tat" Tatlock, "Jonesy" Jones, Chuck Velzy, many others.

Feminine faces, not so familiar but a darn sight more beautiful: Mrs. Harry Schlenz, Mrs. J.J. Wirts, Mrs. C.O. Young, Mrs. N.E. Anderson—and where were the others of the "Lovely Contingent"?

— F.W. GILCREAS, PROG. CHAIR. —

Add another "Cubic Centimeter" to the list, this one Ruchhoft of U.S. Public Health Service.

Don't we all wish we had such a pleasingly deep voice as our Keynoter, A M Rawn?

It was a pleasant sight to see some two dozen or so of Prof. George Barnes' senior sanitary and civil engineering students at the meetings Thursday.

Looks as though the Central States will take the Man-Miles Cup—final count tomorrow.

— CLINTON INGLEE, ENTERTAINMENT —

City of Cleveland Passes are necessary to obtain access to The Easterly Sewage Plant at the inspection trip on Saturday.

Official passes will be issued between 5 PM and 6 PM today, Friday, at the registration desk, to all those having inspection trip tickets. *Be sure to obtain your passes.*

— MORRIS COHN, PUBLICITY —

Pre-convention registration on Wednesday was 76, and to noon, Thursday, was 315; and 359 at 4 P.M.

We weren't kidding about the concert meister, C.C. [Larson], and his symphony. Roy Phillips is a flautist, no less.

— CURRY FORD, REGISTRATION —

What happened to the luncheon on Thursday noon for the NYSSWA group? Perhaps the same thing that happened to the dinner that Stu Coburn owes to Gladys Swope.

Likewise, no one materialized at the Priorities Clinic on Thursday. Don't tell us you don't have a priority problem! Try the clinic from 10 to 12 or from 1 to 3 in the Sec'y's office.

— BUY WAR BONDS & STAMPS —

Did you hear what Commodore J.K. Hoskins said? "No deferment for sewage works operators yet, and no effort by cities to increase salaries in order to hold men."

It used to be "How you 'gonna' keep 'em down on the farm . . .?" Now its "How you 'gonna' keep 'em down at the plant?"

— MRS. J.J. WIRTS, LADIES ENTERTAINMENT —

What to do in an Air Raid

As soon as bombs start dropping
If you are feeling well
Just pay no heed to all the popping
Go out and run like hell

When air raid sirens wail
Be sure to take advantage
If in a bakery, grab some cake—
Who cares about the damage

If in a tavern grab a bottle
Choose brand of which you're fond
If in a show, hold not the throttle
Just pick a seat and grab a blond

If unexploded bomb lies at your feet
As you are passing by
Just dig it out, be very neat
And shake like hell—goodbye!

(with apologies)

— W.J. ORCHARD, FINANCE —

Late stay-er-uppers around the Hotel and elsewhere: Johnnie Horgan, Chan Samson and Al Martin (the Tonawanda Twins), C. George Anderson, Jim Angel. Who else?

Twenty-seven (27) exhibitors! We suggest you take time out from your peripatetic perambulations to look over the exhibits and pick up some ideas as well as some literature.

— L.E. REIN, CHAIR, S.W.M.A. —

We had a bad few minutes Thursday morning before the meeting started when we saw a huge banner in the Ballroom reading "St. Louis in '43"—for our part we want you to "Shuffle off to Buffalo" next.

— ANGUS MacLACHLEN, LOCAL HOST —

If you don't see your name on this page, look on the next.

— D.F. McFEE, W. & S. WKS. MFGS. —

Vol. 3, No. 3   Cleveland, Ohio   Sat. October 24, 1942

Appropriately enough, this war-time conference has been attended by a goodly sprinkling of men in uniform—lieutenants, captains, majors, USPHS commanders—but where is the Navy? Disposing by Dilution?

— PACIFIC FLUSH TANK CO. —

Neatest phrase of the day—Johnnie Johnson's "Modern Four Horsemen of the Apocalypse: Priorities, Rationing, Substitution and Labor-shortage."

— ENGINEERING NEWS RECORDS —

We don't know where you were on Thursday evening, but we had an aisle seat at the smoker. It was neat but not gaudy—and a sight to see the expressions on the faces of the stags while that nifty duo fiddled around the floor.

— CHICAGO PUMP CO. —

You missed something if you didn't hear Mr. Gottschalk and his amazing memory tricks. You missed something also if you left right after the show because he put on some mystifying mathematical tricks for the stragglers who hung around. Wonder if we could get him to calculate our income tax.

And, by the way, have you ever been called a honey-dew melon with legs? But we weren't the only one with a "hole in our hair cut."

— WATER WORKS & SEWERAGE —

And, while the men lunched and smokered, what did the ladies do? Well, if you saw "Jimmie's" [Larson] new hat or that lovely vision in the new gown, Friday night, you will know that the Fashion show was a huge success.

— DORR CO. —

Central states walked off with the 28,220 man-miles cup again this year. Final figures look something like this.

Central States—68 men, 28,200 man-miles

New York—46 men, 16,100 man-miles
California—5 men, 9,700 man-miles
New England—11 men, 5,610 man-miles
Ohio—52 men, 4,160 man-miles
We told you Ohio couldn't win.
Pacific N.W. & Arizona were among those absent.
Total Registration—406.

— SEWAGE WORKS ENGINEERING —

Things I wish I'd said: "Defensive measures and materials have hogged the limelight long enough. We cannot win a war by defense alone so the keynote of this convention will be 'Offensive Materials'."—C.C. Larson.

We don't agree with F.W. Jones' requirements for the Quarter Century Club. Why limit it to men in *charge* of sewage plants 25 years ago?

— WALLACE & TIERNAN CO. —

Thursday, after the session, some very important committees met. Among them the Sewage Works Practice Research and Standard Methods Committee. We think you will be agreeably surprised when the results of these meetings are published, particularly with the new program of the Research Committee.

Friday morning's session evoked some question and comment. We hope you had your priority and civilian defense questions answered.

— PUBLIC WORKS —

One cannot leave this convention without mentioning the excellent talk and reminiscing of Dr. J.V.N. Dorr—a scholar, philosopher, scientist, and gentleman.

— RALPH B. CARTER —

We shall never forget Frank Lovett collecting 15¢ from Harry Schlenz' dinner guests to pay for the roast beef and then winding up with the check. Frank has some good stories, too, but ask *him* what they are.

— CHAIN BELT CO. —

Unfortunately, this issue had to go to press before the symposium on Friday afternoon. We know that several of you attended the symposiums on Thursday night in Rooms 714, 452, 844, 1250, and others.

We did catch the opening on the Friday P.M. session and it started off with the admonition not to pull punches although abiding to the Marquis of Queensbury rules.

— LINK BELT CO. —

When this issue hits the stands on Saturday morning, we shall be attending the Operator's "Turn" Table and in the afternoon the inspection trip.

We have enjoyed the whole convention, we hope you have *also*. It's been a pleasure in more ways than one.

A year from now is twelve months away and who knows—we perhaps can only hope to see you all again. In Buffalo? ? ? ?

And that'll be all for a spell!

Sorry we couldn't mention more of the fine exhibits.　　G.E.S.

THE FEDERATION DAILY

GEORGE E. SYMONS, Editor

PUBLISHED
for the
SEWAGE WORKS
FEDERATION
by
WATER WORKS
AND
SEWERAGE

FOURTH
ANNUAL MEETING
of the
SEWAGE WORKS
FEDERATION
CHICAGO
OCTOBER 21-23
1943

Vol. 4, No. 1   October 21, 1943

SCOOP!!!   *EXTRA!!!*

This is not on your program . . .
Hon. Maury Maverick, Dir.,
Governmental Agency WPB
will appear on the Program . . .
Friday Morning 9 a.m.

HEAR ABOUT . . . P-141

From the Chemical Preview, we learn that red headed persons have fewer hairs than brunettes or blonds, the averages being 90,000, 122,000, and 140,000. Some of us have even less than 90,000. —"A Honey Dew Melon with legs"— INDEED! . . . We're back—have a room and how are you?

It's news, if you haven't heard it! Stan Barker of New York is now a Commander in the Navy and stationed at Lima, Peru in South America! . . . Our own Larry Lingor (WW&S) in the Navy since August 31st, completed his "Boot Training" at Great Lakes, this week!

We had a letter from Lt. Col. Arthur Morrill on Monday. He used to be at Detroit, remember? Now located in India. Art writes: "I have been having a very interesting time. For five months I worked on general camp sanitation, including mosquito control. In the latter I now talk glibly, among the uninformed, about Anopheles minimus and distinguish male from female mosquitoes at five feet. Most of the water plant superintendents never heard of pH and one at least thought it meant 'permanent hardness.' In India the cow is sacred, but not the purity of water supplies."

Plug—Eimco is showing Movies—all other advertising accepted at regular rates.

Ask G.E. Flowers of the Southerly Plant of Cleveland about Saran String for filter cloth string substitute. Costs three to four times as much but lasts more than 25 times as long.

Program Correction: Friday afternoon—Central States Annual Meeting will be held in Bal Tabarin . . . NOT in the Grand Ballroom!

You can blame the Truckers' Association for the lack of rooms available. They promised to release 400 rooms Wednesday noon, but released only about 25 by Wednesday night. . . . No legal recourse says Harry Schlenz.—Maybe it's a good excuse for the situation but we know a lot of fellows who still don't like it.

Vol. 4, No. 2   Fri., October 22, 1943

Registration to 5 p.m. Thurs. Oct. 21 was an even 514. Not bad at all. Last year boasted but a neat 470 according to Secy. Wisely. 39 Ladies this year.

Well now you know! That is if you heard Ed Cleary Thursday—you know why he shunned the beer at the Smoker. When one only has two good drinks of water in 16,000 miles, one gets rather tired of beer.

Were we surprised—only the Chicago Pump Co. placed an advertisement with us. Something about files on operation records. But like all papers, we are not accepting advertising, because of the paper shortage.

Exhibitors this year totaled 35 and utilized 48 booths, according to Arthur Clark, Secy. Mgr. WSWMA, who has a booth at the head of the Exhibitors Floor. If you are a manufacturers representative, you should meet Art. He's not hard to find—he too has something less than 90,000 hairs on his head. Art had to go to Schleins (not Schlenz) for dinner recently to hear about a fire at home!

If you want a new blue suit . . . sanitary engineers are wanted by the CEC of the U.S. Navy for service with the Superintending Civil Engineer in each of 7 Naval Districts. See Comdr. H.O. Lord, CEC USN (R), Office of Superintending Civil Engr., 938 Bd. of Trade Bldg., LaSalle and Jackson Sts., Chicago.

The program committee had a red face last night because that extraordinary picture about "Big Inch" didn't arrive. Half of it did and the other half is expected yet. If it arrives, you'll see the picture sometime on the program before Sat. noon. For the smoker we are all indebted to Tom Wolf of the Cast Iron Pipe Research for coming through with the movie on water pollution.

Things we'll never know . . . How come Joe Gilbert and Rolf Eliassen had women's badges—why Frank Roe has kept us in the dark for two months about young Pete Roe (someone is sure to nickname him Shad)—and why we didn't have a flash gun to get that picture of the Red Head and a certain Editor at the Smoker.

The Grease Committee made a decision last night on how to determine Grease. Ask Hap Hatfield.

This isn't original with us—Neither was it original with the fellow who put it in the box located in booth 27—but it's good . . .

"A Coordinator is a man who brings organized chaos out of regimented confusion."

"A Conference is a group of men who, individually, can do nothing but as a group can meet and decide that nothing can be done."

"A Statistician is a man who draws a mathematically precise line from an unwarranted assumption to a foregone conclusion."

"An efficiency expert is a man who knows less about your business than you do and gets paid more for telling you how to run it, than you could possibly make out of it even if you ran it right instead of the way he tells you to."

"And, a Consultant is a draftsman away from home."

Vol. 4, No. 3   **October 23, 1943**

What a Convention . . . 604 registrations at least, as we go to press! And, the information you have been waiting for is that Central States again won the man-miles contest. First four leaders:

Central States	192 members	27,070 miles
New York	40 members	25,240 miles
Pennsylvania	17 members	8,600 miles
Ohio	23 members	3,700 miles

As a New Yorker, we're getting rather fed up on being in second place all the time. What have those Central States boys got? They even have the most members now with 495 to New York's approximately 478. How about that, Mr. Bedell?

Contributed:

Smoke Screen News: "A certain gentleman at the speakers platform is requested to refrain from incineration practice while speakers are expounding!" . . . Do you know who it is? We don't.

Contributed:

"When Hitler and Tojo have felt the knife,
That will bring the end to Global Strife,
Blueprints will harness all sewage flow,
Then A.W.O.L. the world will know . . . means,
    American Way of Life"

We stopped at one table at the Banquet where there were six past presidents of something or another entertaining one charming lady. Will the six please take a bow!

Arthur Akers of Public Works wants an emphatic denial that he was the man who started the bonfire in the middle of State and Madison Streets at 2 a.m. on Friday morning. It was two other fellows!

During one of the social hours we found most of the U.S. Public Health Service in the J-M room. (Is that enough of a plug for you fellows?)

Quip of the day: Morris Cohn said "Detroit is the city where sewers as well as Sedans are built." And while we're referring to Editors, Bill Foster (American City) says we spelled his name wrong in the list. . . . Have you ever tried to run a mimeograph stencil at three or four a.m.? Please forgive the slight errors, but some of you fellows should learn to write so one can read it.

An Ohioan says that he feels like a native after two days here in Chicago, for when he took a three block walk, four persons stopped him to ask directions!

Casey Jones blew in . . . or did he say flew in . . . early Friday morning!

Of course you all attended the Banquet and heard that "Pete" Wisely is going to be the new full-time Secretary, Editor and Advertising Manager of the Federation. A triple threat man . . . but who else? And you know that the next year's meeting is in Pittsburgh, and that several awards were made by the Federation as follows:

Kenneth Allen Awards

A.H. Wieters	Iowa
R.S. Phillips	No. Carolina
J.K. Hoskins	Federal
E.F. Eldridge	Michigan
W.H. Wisely	Central States

George B. Gascoigne Award: Lt. Kerwin L. Mick
Charles A. Emerson Award: Dr. Floyd W. Mohlman
Honorary Membership-Federation: Charles Gilman Hyde
Honorary Membership-Federation: H.E. Moses
Honorary Membership-Central States: Langdon Pearse

Thank Harry Schlenz for some swell entertainment at the Smoker and the Banquet. And if you didn't hear the Four Harmonizers (National Champions of the "Society for the Promotion of Barber Shop Quartet Singing in America") in our Room 641 . . . you really missed something.

That was all swell, but the best entertainment we heard or saw were three Australian fliers who had just completed training . . . singing, "Waltzing Matilda." We don't think it will comfort the enemy to know that before long they will be in England. Pardon us while we take off our hat in salute and wipe a tear.

If you haven't sat in on Frank Lovett's stories you have missed a treat. He and Mr. Veatch are now telling the same one about the man naming the three puppies. The first—Clark Gable, because he was the handsomest . . . the second—Joe Louis, because he was the scrappiest, and the third—Paderewski, because .. .. We don't get it!

Bart Marshall said it . . . "This is the best convention the Federation has had." We echo!

There are a lot of other things we thought we would say before we signed off this year. We regret that we can not publish the entire registration list at once in this issue . . . but, the nights in Chicago just haven't been long enough to do everything!

We've enjoyed being "HERE" and hope you see us "THERE" in Pittsburgh . . . next year! ! ! ! !

WPCF Awards and Medals

The mechanism for making Federation awards was developed between 1941, when awards were recommended by the General Policy Committee, and 1943, when the first awards were made. These included the Harrison Prescott Eddy Medal, George Bradley Gascoigne Medal, Charles Alvin Emerson Medal, and Kenneth Allen Awards. Awardees were selected by an Awards Committee which was established in accordance with a recommendation of the General Policy Committee and approved at the 1941 Board meeting, and based on recommendations of this Awards Committee and approved by the Board at its 1942 meeting. From time to time, the Board authorized additional Medals and Awards. Descriptions are provided for each that are currently being awarded. These descriptions, based on publication in various issues of the *Journal*, provide information on the origin of the award, its purpose, and criteria for selecting awardees. They are arranged in alphabetical order by name of the award. With the exception of the Bedell Awards, all Federation Awards are presented to the recipients at the Annual Conference banquet. All awards are approved by the Board of Control.

Citation of Excellence in Advertising

This award was authorized by the Board of Control on September 27, 1964. It is given in recognition of outstanding contribution to the water pollution control field through the effective and original presentation of equipment and processes in a display advertisement appearing in the cited issue of the Federation *Journal*. The first awards were made in 1966.

Judging is done by a special committee appointed by the President of the Federation. This committee is a subcommittee of the Awards Committee. An award is made for each of the three categories: black-and-white, two-color ads, and three-or-more color ads (inserts). Honorable mention certificates may be awarded to advertisers in any of the three categories.

All entries must be received by April 1 following the year for which the award is to be considered.

Kenneth Allen Award

This award was created by the Board of Control in 1943 for the purpose of recognizing persons for outstanding service in sewerage and sewage treatment works field, as related particularly to the problems and activities of any member association. This award was discontinued after 1948 as a result of a request by the New York State Sewage Works Association which had an award with the same name that predated the Federation's award. Beginning in 1949 this award was succeeded by the Arthur Sidney Bedell Award.

Arthur Sidney Bedell Award

To acknowledge extraordinary personal service to the member associations, the Arthur Sidney Bedell Award was established in 1948. Each Member Association is privileged to name one of its members to receive this award, which may be based on organizational leadership, administrative service, membership activity, stimulation of technical functions, or similar participation. The frequency of nomination varies from annually to once in three years, depending on membership. The award is named for the second president of the Federation, who exemplified its purpose by his long devotion and service to the New York WPCA. Certificates are presented to the awardees at the member association meeting following Board approval of the individuals.

By action of the Board of Control on October 14, 1951, criteria for selecting awardees were changed. These can be found in the March 1952 issue, Part II of the *Journal* under the Awards Section. Awardees are selected by the respective Member Associations and approved by the Board of Control.

Thomas R. Camp Medal

The Thomas R. Camp Medal, established by the Board of Control on October 1, 1964, is awarded annually to a member of any Member Association of the Federation who has demonstrated, by design or the development of a wastewater collection or treatment system, the unique application of basic research or fundamental principles. The award honors Thomas R. Camp, founder of the consulting engineering firm of Camp Dresser and McKee, Boston, Mass., who was an outstanding consultant, educator, and technical author, and who made many notable contributions to the water pollution control field. The first award was presented in 1965.

Awardees are nominated by a subcommittee of the Awards Committee.

Harrison Prescott Eddy Medal

The Harrison Prescott Eddy Medal is awarded annually to a member of any member association of the Federation for outstanding research contributing in an important degree to the existing knowledge of the fundamental principles or processes of wastewater treatment, as comprehensively described and published during any stated year in the Federation *Journal*. The award commemorates Harrison Prescott Eddy, a famous engineer and a pioneer in the art of wastewater treatment. The first award was presented in 1943, having been approved by the Board of Control in 1942. Awardees are nominated by a subcommittee of the Awards Committee.

Charles Alvin Emerson Medal

The Charles Alvin Emerson Medal is awarded annually to a member of any Member Association of the Federation for outstanding service in the collection and treatment of wastewater, as related particularly to the problems and activities of the Water Pollution Control Federation in such terms as the stimulation of membership, improving standards of operational accomplishments, fostering fundamental research, etc. This award honors Charles Alvin Emerson, who was president of the Federation from 1928 to 1941 and who became its first honorary member. It was first presented in 1943, having been approved by the Board of Control in 1942. Awardees are nominated by a subcommittee of the Awards Committee.

Gordon Maskew Fair Medal

The Gordon Maskew Fair Medal was created by the Board of Control in 1967. This award serves the area of engineering education in the field of water pollution control. As professor of sanitary engineering at Harvard University, Cambridge, Massachusetts, Gordon Maskew Fair did noble work in preparing students for the professional field of sanitary engineering and professional positions. This medal may be awarded annually to a Federation member for proficient accomplishment in the training and development of engineers, particularly at the graduate level. It was first presented in 1968.

In the background of the establishment of the award, the minutes of the Executive Committee show that it was the intent, after sufficient awards had been made, that a subcommittee of the Awards Committee would be established. Until that time the Executive Committee would act as the stimulating group and the Award Committee. Since only three awardees have been chosen and only two of those are living, an awards committee had not been established by 1976.

George Bradley Gascoigne Medal

The George Bradley Gascoigne Medal is awarded annually to a member of any Member Association of the Federation for outstanding contribution to the art of wastewater treatment plant operation through the successful solution of important and complicated operational problems, as comprehensively described and published during any stated year in the Federation *Journal*. This award is in memory of George Bradley Gascoigne, a prominent consultant from 1922 to 1940, who demonstrated an unusual interest in the operation of wastewater treatment plants. It was first presented in 1943, having been approved by the Board of Control in 1943. Awardees are nominated by a subcommittee of the Awards Committee.

William D. Hatfield Award

For the years 1946 through 1954 the William D. Hatfield Award recognized outstanding annual reports on wastewater treatment plant operation. After the first year, awards were made on the basis of plants serving (1) less than 10,000 population, (2) populations of 10,000 to 100,000 and (3) populations of more than 100,000.

This award was revised so that since 1956 it has been given for outstanding treatment plant operation. At that time, the frequency of this nomination was changed to correspond with the Bedell Award, which is at intervals of one, two, or three years, depending on the number of members in the Member Association. Certificates are presented to the awardees at the Member Association meeting following Board approval of the individuals.

Criteria and suggestions for selection of awardees were revised by the Board in 1949 and in 1956. The current rules are found in the *Journal's* March issue, Part II. Awardees are selected by the respective Member Associations.

Philip F. Morgan Medal

The Philip F. Morgan Medal was established by the Board of Control on October 10, 1963. The award is made to a member of any Member Association of the Federation for the in-plant study and solution of an operating problem; publication of a paper is not required. The criteria include originality, significance, comprehensiveness, effort, and most importantly, the verification of an idea. Two award classifications are: (a) plants serving more than 5,000 population, and (b) plants serving less than 5,000 population. This award honors Philip F. Morgan, who served with distinction as professor of sanitary engineering at the State University of Iowa from 1948 to 1961. An outstanding practical researcher, he maintained a strong interest in plant operation. Certificates of Merit may be awarded in addition to the primary award of a plaque. It was first presented in 1965. Awardees are selected by a subcommittee of the Awards Committee.

William J. Orchard Medal

The William J. Orchard Medal was established by the Executive Committee, December 14, 1960. The award is made for distinguished service to the Water Pollution Control Federation and is to be given as considered appropriate by the Executive Committee. Recipients of the award are recognized at the annual banquet by the presentation of a distinctive, engraved, bronze-on-walnut plaque.

The award was kept secret by the Executive Committee from December 1960 until it was first presented to W.J. Orchard at the 1961 Federation Conference in Milwaukee. It has been presented to only four other persons: Gordon M. Fair, 1964; Harry E. Schlenz, 1968; Harris F. Seidel, 1971; and Frank H. Miller, 1974.

Willem Rudolfs Medal

The Willem Rudolfs Medal was established in 1949 as the Industrial Wastes Medal to be awarded annually in the form of a plaque to a member of any Member Association of the Federation for the most outstanding contribution by an industrial employee on any aspect of industrial wastes control, as published in the Federation *Journal* during the year preceding the award. The award was renamed in honor of Willem Rudolfs by Board of Control action in 1966. It was first presented in 1950.

Member Association Safety Award

The Member Association Safety Award, established in 1970, is designed to stimulate Member Associations to conduct vigorous safety programs in local wastewater works and to encourage the collection of injury statistics on a national basis. Requests for consideration for this award are submitted by Member Associations, together with supporting exhibits to document and illustrate the association's safety program. Important factors considered by the judges in making the award include (a) the Member Association safety program, (b) cooperation on safety with other organizations, (c) safety publicity, (d) safety materials and visual aids used, (e) collection and use of injury data, and (f) wastewater systems personnel injury experience for the past five years in the area served by the Member Association.

Based on information submitted by the Member Associations, a subcommittee of the Awards Committee recommends awardees for the approval of the Board of Control. Certificates of the award are presented at the annual conference of the Federation. It was first presented in 1970.

Harry E. Schlenz Medal

The Harry E. Schlenz Medal was established by the Board of Control in October 1970. It may be awarded annually for distinguished service in promoting public awareness, understanding, action in water pollution control and may go only to someone whose principal employment is outside the technical field. The award is in memory of Harry E. Schlenz, who served for many years as President of the Pacific Flush Tank Company and as Federation President in 1961 62. Mr. Schlenz was a distinguished leader in promoting public understanding of the need for adequate water pollution control. The award was first presented in 1971. Awardees are nominated by a subcommittee of the Awards Committee.

Honorary Membership

Honorary Memberships are proposed annually to the Board by a Constitutional Committee for the purpose. Criteria and rules for nominating and electing honorary members are found in Section 2.46 of the Federation bylaws. At the present time, living honorary members may not exceed a ratio of 1 per 350 active members and no more than three persons may be elected in any one calendar year. A negative vote of 10 percent is sufficient to reject a nominee. The first honorary member was elected in 1941. With few exceptions, additional members have been elected annually.

Collection System Award

The Board of Control at its 1973 meeting in Cleveland authorized a new award to be known as the Collection System Award. The Executive Committee adopted guidelines and procedures to be used in selection of the recipients for this award. The first award was made at the Denver 1974 meeting of the Federation.

"Thanks to the Chemical Foundation"*

This issue of the *Journal* is notable for the fact that it is the first issue to be produced without some degree of participation by The Chemical Foundation of New York City, the institution which brought the Federation and the *Journal* into being. In view of the intensely important service rendered by Chemical Foundation and its modest, "behind-the-scene" role, a suitable acknowledgement is very much in order. . . .

The generosity of the Foundation's original subsidy is apparent when the set-up of the first business office of the *Journal* is considered. Office facilities were provided in the Foundation headquarters and Mr. Buffum assumed the duties and responsibilities of Business Manager of the *Journal*. These duties included solicitation of subscriptions, solicitation and production of advertising, maintenance of mailing lists, and financial accounts, and printing and mailing of the *Journal*. Finally, and of utmost importance, the Foundation undertook the obligation of meeting any deficit incurred in the entire venture. It may be of interest that the original agreement covered a period of but one year; yet Mr. Buffum continued to serve as Business Manager from May, 1928 to the time of his death in June, 1940, while the Foundation extended its participation to January, 1944.

At the death of Mr. Buffum in 1940, Dr. Rudolfs wrote fittingly of the work Mr. Buffum had performed in behalf of the Federation and the *Journal* [**12**, 816 (1940)]:

> "Those who have come in contact with Mr. Buffum can appreciate his work and encouragement to stimulate the development and research in sewage treatment. His identification with the *Journal* made it financially possible to reach a large number of sewage plant operators. His interest was an important factor in the rapid growth of sewage research, which had made this country the leader in the world. He was always ready to support any movement to fight diseases and secure cleaner streams. In the earlier years of the Federation his advice and help was an important factor in the rapid growth of Sewage Works Associations in this country. His business-like manner in handling the affairs of the *Journal*, his sympathy and support carried the *Journal* over several difficult periods."

The Foundation continued its service to the Federation and the *Journal* after Mr. Buffum's death, designating Mr. Arthur A. Clay to serve as Business Manager. Mr. Clay worked zealously and enthusiastically in the position and was particularly successful in giving impetus to the use of advertising facilities offered by the *Journal*. The first Convention Number was attempted at his suggestion and, under his guidance, proved to be an outstanding success.

In January 1941, after the Chemical Foundation had carried the *Journal* through its first twelve years of existence, the opening move was made to relieve the Foundation of this burden. It was then that the Federation Secretarial office was expanded to assume the bulk of the responsibilities

---

* An editorial in the Feb., 1944 issue of *Sewage Works Journal*.

of business management, but the Foundation was still retained to provide assistance in the mechanical production of *Journal* advertising. Mr. Clay's title was changed to Advertising Manager and the work was carried on under his supervision by Miss Gertrude R. Horan, his capable and efficient assistant. It is significant that more than 200 pages of paid advertising were carried in the *Journal* in 1943, an all-time high.

At best, the writer can but echo the expressions of gratitude to The Chemical Foundation which are in the thoughts of Messrs. Emerson, Mohlman, Orchard, Rudolfs, and others who have been so intimately familiar with the part taken by that agency in the development of the Federation. The writer can and does offer sincere thanks in his own right, however, for the courteous and complete cooperation that he has enjoyed during the past three years in his gradual assumption of the functions formerly provided so efficiently and unostentatiously by the Foundation. . . .

Chemical Foundation, the Federation thanks you, deeply and sincerely!

W.H.W.

FSWA Special Report on Research Activity
January, 1953

On October 9, 1952, the Board of Control entertained a proposal that the Federation give greater emphasis to the encouragement and stimulation of research in its areas of interest. This proposal recommended expansion of the headquarters staff by the employment of a qualified research man, if that should appear to be desirable and necessary.

The Board directed that the matter be referred to the Executive Committee with the chairmen of the Finance and Research Committees for study and specific recommendation. This was done under date of October 22, 1952. Several constructive letters were exchanged among the committee personnel involved, but response to date has been incomplete and inconclusive.

It is intended that this report will provide a basic analysis of the proposal that will expedite its further consideration.

Research Functions of the Federation

It is assumed initially that the Federation will not actually operate research facilities, and that this is not to be considered an ultimate objective. How, then, can the Federation best serve in this area? Some possible activities are listed herewith:

1. Development and maintenance of a directory of agencies and institutions engaging in research studies pertinent to the Federation's interests. Data might be included on staff, facilities, and the type of work that might be specialized. The directory should include research laboratories in educational institutions, governmental agencies, trade associations (National Canners Assn., etc.), and those that are privately owned or endowed (Mellon Institute, etc.).

2. Surveys—probably annual—of research that is in progress on problems within the Federation's realm of interest, with data to be distributed to all research agencies.

3. Analysis of the needs for research in the field, and the outlining of specific problems on which fundamental information is needed.

4. The "selling" of needed research projects to qualified research agencies that are best equipped to undertake specific studies.

5. Maintenance of advisory contact with all research agencies, to prevent duplication of effort and to serve as a clearing house for interchange of information.

6. Publication of research papers and reports in the *Journal* or in special bulletins.

7. Reviews of the research literature, both domestic and foreign. General reviews should be at least annual in frequency. Special reviews on specific topics should be done as needed.

Past and Present Research Activities in the Federation

To date, research functions in the Federation have centered mainly in the Research Committee, with limited general activity in the headquarters office.

The Research Committee has produced splendid annual reviews of the research literature since 1932. Special reviews on toxicity of sewage and wastes have been published and are being expanded. Surveys of research then underway on water pollution and wastes disposal were made and published in 1943, 1944, and 1945.

Publication of research papers and reports has been featured in the Federation *Journal* from its inception. It is without doubt the leading medium for the exchange of such information.

The headquarters office has had occasional opportunity to afford assistance and guidance to research agencies. Some research projects have originated as a result of editorial reference in the *Journal* to knowledge that is needed.

From 1947 to 1952 the Federation sponsored an extensive research project on analytical methods, which was financed by the National Institutes of Health. This work was directed by the Committee on Standard Methods, and administered by the headquarters office.

Obviously, there is much more that could be done toward the objectives outlined previously.

Other Agencies Concerned with Research Activities

The Federal Water Pollution Control Division is interested in nation-wide research, and has authority to carry out a program similar to that which has been outlined for the Federation. Its plans in this direction are unknown.

It is reported that the National Science Foundation has established a scientific information office for service as a clearing house for guidance of fundamental research.

It is not suggested that the Federation should sidestep any responsibility for assumption by these or any other agencies, but the possibility of conflict and overlap merits consideration.

Possible Plans of Action

Herewith are listed several plans that might be followed by the Federation in emphasizing its research activity. No doubt there are others, and it is also likely that one or more of the following could be combined to advantage.

I. Reorganize and expand the present Research Committee, creating subcommittees to undertake separate objectives as outlined. An annual appropriation of about $1,500 would probably cover necessary additional travel and additional headquarters office expense.

II. Employ a part-time research director (or possibly appoint the Research Committee chairman on a part-time basis), with procedure generally as stated in Plan I. Annual cost probably $3,000 to $5,000.

III. Employ a full-time administrative assistant to the executive secretary, to enable the headquarters office to assume additional research functions in its administrative routine. This plan would retain the Research Committee on its present basis. Mr. Orchard suggests (his letter of January 12, 1953) an annual appropriation of $8,000 to $10,000 to be necessary.

(Note—The headquarters office probably could undertake annual questionnaire surveys of research institutions and projects with its present personnel, with some assistance from the Research Committee.)

IV. Employ a well-known leader in water pollution control research as director of research, to assume full supervision and execution of the entire program. This is essentially the solution suggested by Dr. Heukelekian in his letter of October 27, 1952. Probable annual cost $12,000 to $15,000.

Closing Statement

It is hoped that this brief summary will be of assistance to the Executive, Finance, and Research Committee personnel who have been requested by the Board to offer recommendations for action.

Respectfully submitted,
W.H. Wisely, Executive Secretary

Script for
"This is Your Life"
Ted Moses Night
Joint Meeting-Penna. Water Works Operators Assn. and the
Penna. Sewage & Industrial Wastes Assn.
8:30 P.M.
August 23, 1955
Hetzel Union Building
Pennsylvania State University
State College, Penna.

(Following the banquet, the introduction of persons at the head table, and the presentation of the awards, Prof. R.R. Kountz will take over at the speaker's table. The lights will be lowered and a spotlight turned on Prof. Kountz.)

Prof. Kountz: Ladies and gentlemen: At this time I'd like to call your attention to an omission in the program. It does not list any entertainment. Actually, we have arranged for entertainment tonight. In fact, we have imported an entertainer—imported all the way from Larchmont in Westchester County, New York—and here he is, Dr. George E. Symons, better known to all of us simply as "Doc."

(During this introduction, Symons makes his way unobtrusively to the steps of the stage behind the head table. At the mention of his name, the spotlight shifts to him and he steps to the microphone.)

Dr. Symons: Good evening, Ladies and Gentlemen, here's that man again! And while this may be a surprise to you, it is not to me. And before I go any further, I want to ask that no one leave this room for the next three minutes. After that I dare you to leave.

(At this point, Symons ad libs a few remarks with reference to the flood, Prof. Kountz' introductory remarks and some "gags" to throw the audience off of any possible inkling as to the true nature of the program.)

Symons: Inasmuch as the program doesn't tell you what I'm going to do to entertain you, perhaps I should say that some months ago, the committee asked me to show you a few slides and to make such remarks as might be pertinent. So without further ado, may I have the first slide please.

(Slide No. 1—picture of an old Penna. house. While slide is being shown, spotlight remain on Symons.)

Symons: My story begins in a little house like this in Muhlenberg, Luzerne County, Penna. It is 1875 and a young Methodist minister and his wife have just become the proud parents of a baby boy....

(At this point, Symons picks up the book and says):

... This is your Life ...

(As he says "This is your Life," a spotlight is turned on the head table and directly on Mr. Moses)

... Howard Eugene (Ted) Moses!

(The lights go on, the audience stands and applauds, led by the committee; Roy Kountz then escorts Mr. Moses up to the stage, and stations him at the microphone.)

GES: Yes! Ted Moses, This is Your Life—and now you know why I met with Deac Matter, Bernie Bush, Al Estrada and Roy Kountz in Harrisburg in May. And I guess you and the audience both know why there wasn't any announcement of this in the program.

GES: And now to get on with my slides . . .

(The lights go out and slide No. 1 comes on again and the spotlight is turned on Mr. Moses and Dr. Symons while all subsequent slides are shown)

. . . Ted, this is not the actual house where you were born—we couldn't get a picture of that house—but it must have been something like this.

Ted: Moses was born down in the bulrushes—or so I believe the Bible says.

(Slide 2—Altoona High School)

GES: Like all minister's families you moved around a lot as a boy, but you settled in Altoona, Pa. long enough to attend this high school.

(Slide 3—Ted Moses in high school)

Ted: I remember that!

GES: It is 1894 and here you are ready to graduate from Altoona High School after only three years. This you accomplish by doubling up your studies, including Greek, in your third year so you would have sufficient credits to enter college.

(Slide 4—Ted Moses in college)

GES: Now it is 1898—You are about to graduate from Dickinson College in Carlisle, Penna. You started out in a liberal arts course but in your senior year you were aiming to become a physician. You were a typical college boy, you belonged to Beta Theta Pi Fraternity and you sang baritone in the College Glee Club and in a free lance quartet.

You left Dickinson College with a diploma under your arm and enrolled at the University of Maryland where you studied dentistry for one year and found it not to your liking.

(Slide 5—picture of F & M Works)

GES: It is now 1900 and with thoughts of medicine and dentistry behind you, you take a job as a draftsman with the Harrisburg Foundry and Machine Works . . .

(Slides 6 and 7—pictures of F & M Works)

You stayed with the F & M Works for eight years and ended up as a steam engine designer.

Ted: I haven't any secrets at all! I don't know where you got all these things.

GES: Two years after you went to work in Harrisburg—it is 1902—you are married to Jane Beckley of Bloomsburg, Pa. . . .

(Slide 8, baby picture of Gene Moses Wilkinson)

And to this union a daughter Gene. In 1908, you joined the Penna. Health Dept. as an assistant engineer—and you have remained with that department for 47 years, rising to the position of Chief Engineer in 1937 and later to Director of the Bureau of Sanitary Engineering, a position you still hold.

You played an important part in the early days of the department. Dr. Samuel G. Dixon, Pennsylvania's first commissioner of Health and his Chief Engineer, Herbert P. Snow, soon recognized your ability as an organizer and you were placed in charge of the department's investigations of major typhoid fever epidemics including Pittsburgh, New Castle, and Erie . . .

(Slide 9—Bayliss Paper Co. Dam)

It is 1910, and you are directing the relief work following the failure of the Bayliss Paper Company's dam at Austin in Potter County, where a large number of people lost their lives and a part of the town was destroyed. Do you recognize that dam, Ted?

Ted: Yes.

GES: Not all of your work was so grim or depressing. In 1921 you attended a short course in Public Health at the Johns Hopkins University in Baltimore; with you were four other engineers from the department including Chief Engineer Charlie Emerson, and Engineers C.B. Mark, I.M. Glace, and G.D. Andrews. And while there, you and another fellow cooked up a trick on one of the party. The story went like this.

(From behind the screen, Mr. C.B. Mark speaks into the backstage microphone.)

Coleman B. Mark: Upon our arrival in Baltimore, Mike Glace who drinks only water, commented on the chlorine taste, but I had coached all the others to claim they could not detect chlorine in the water. The argument continued all week and Glace announced that he intended to bring his orthotolidine testing kit to Baltimore the following week to prove that the Baltimore Water was highly chlorinated.

I telephoned Glace's assistant in Harrisburg and instructed him to replace the orthotolidine in Mike's test kit with water before Mike came home for the weekend. This was done and when Mike Glace made the test before our group the following week, no chlorine residual was indicated. Do you remember, Ted?

Ted: I do indeed—Coleman Mark—how did they get you out of Washington?

(The curtains part and Coleman Mark steps out to shake hands with Mr. Moses)

GES: Yes, Ted, this is Coleman B. Mark, long time friend, who served with you in the department for 26 years, and who served the State of Pennsylvania for 33 years and gave 28 years to the U.S. Army and National Guard, rising to a Lt. Colonelcy in the Army of the U.S. during World War II. Col. Mark is now President of the Engineers' Club of Washington, D.C.

Ted: This is a surprise, I'm sure.

Mark: I want to take this opportunity before all these engineer friends of mine and yours to thank you for the guidance and inspiration of work you gave to me during my engineering career. This is indeed a happy occasion.

Ted: That's nice of you! If you hadn't been a good fellow, I wouldn't have been any guidance to you, but it was a nice time we had together, I'm sure.

Mark: Congratulations on all these Honors and Best Wishes to you.

Ted: Thank You! I'm glad to see you again.

GES: When Ralph Edwards gave us permission to do this program for you Ted, he suggested that we follow your life in sequence, but Hazel Bishop who sponsors "This is Your Life" is not sponsoring our show tonight so I'm going to take the liberty of bringing in something out of sequence here.

When Deac Matter and Bernie Bush were together in the Wilkes-Barre District office, you used to go up to Wayne County frequently to fish with them. You always stopped to pick up the fourth member of the party.
(From behind the screen, Frank Scheurholz speaks.)

Frank Scheurholz: This cigar store of mine also sold fishing tackle and served as a meeting place for many of Honedale's male citizens. I didn't have any clerks, but when I went fishing, the store operated as usual. My customers waited on themselves and deposited the money in the cash register. At night if I hadn't returned, the last man in the store locked up.

GES: Who is it, Ted?

Ted: Sherry, isn't it?

(The curtain parts and out steps Mr. Scheurholz.)

Ted: Well, well, Sherry, I'm glad to see you!

GES: Yes, Ted—Mr. Frank (Sherry) Scheurholz, Honedale cigar store proprietor, extraordinary, and your long time fishing companion.

Ted: I couldn't forget that entirely. We were up on a lake there and he and another friend of mine were over in one boat by themselves and they had a bottle between them and they had a rule that, every time they caught a fish, they had to take a drink but they were catching them so fast they were emptying the bottle, so they had to cut it down for a drink for every two fish. Do you remember that Sherry? It's a great pleasure to see you again. How are the chickens? He's a chicken fancier.

Sherry: I just want to say the reason we caught so many fish is that bottle had to be passed between the other fellow rowing the boat and me and after a few shots, I figured the boat was getting pretty heavy so I dropped it in the lake. It just goes to show if you have the right kind of bait, especially liquid, you can catch the fish.

Ted: Thanks a lot, Sherry.

Sherry: Congratulations, Ted and don't forget that life begins at 80.

Ted: That's right! Pretty close, now!

GES: And now let's go back to your work. In 1922 you became As-

sistant Chief Engineer and in 1937 when you became Chief Engineer the state legislature passed the first stream pollution control act, which made possible the control program in Pennsylvania. Meanwhile you always found time to play a practical joke on some of your men.
(At this point, a voice behind the screen speaks.)

Glace: Mr. Moses, I'm calling you from Schwenkville, in the lower part of Berks County, below Boyertown, and I'm the superintendent of the municipal water works. We had a terrible rain down here and it washed away our standpipe. Your man and District Engineer Andrews were down here recently and approved that standpipe and now it's completely gone and an old lady drowned in the stream. We want Mr. Andrews to come down here and tell us what to do.

GES: Do you remember the incident, Ted?

Ted: I remember that distinctly—Well Mike Glace was in on it and Harry Freeburn. It sounds like Mike Glace.

GES: Yes, Ted—here he is—the one, the one and only, Ivan Maxwell (Mike) Glace, who with your blessing and help, organized the Penna. Water Works Operators Assn.

Mike: That was part of the deal for getting after the gang who pulled the job in Baltimore, which was only a year or so later. I went in to tell Ted that it was pretty smart of me calling from the extension phone outside. We had Andrews all excited up to the point where George climbed into his car to start for Schwenkville when Matter stopped him. Do you remember that, Ted?

Ted: I do indeed but I'd say they got back at you when they pulled the orthotolidine trick on you.

Mike: No, I think you're wrong! The orthotolidine trick was first.

Ted: No, it was after!

Mike: I congratulate you anyhow Ted!

Ted: Oh! Thank you a lot Mike.

Mike: It's nice to have this for you and I'm just delighted to be here. You know it was just 30 years since we came up here for the first time. Remember the old group of filter plant operators and sewage plant operators that came up when we organized.

Ted: That's right!

Mike: That was in 1925.

Ted: That was the beginning.

Mike: That was before we organized the Filter Plant Operators' Association.

Ted: We must have lived right; we had a good growth.

Mike: Congratulations!

Ted: Thank you! No secrets left, that's certain.

GES: You were quite an organizer of new associations and new societies. You helped organize the Penna. Sew. & Ind. Waste Assn. and the Federation, and one very famous society. In 1937, you conceived the idea of honoring, in jest, three or four persons each year by inducting them, as Knights of the High Hat, into the Sludge Shovelers Society. Fifty-five times in 15 out of the past 19 years, you have said, "By virtue of the authority invested in me, I now crown you with the High Hat. You will repeat after me, three times, the words Sludge Shovelers Society." Believe me, those of us who have been so crowned are mighty proud of the honor, and proud to wear the symbol of the shovel.

It is now October 1951, in St. Paul, Minn. The Federation banquet is over, and you are called to the head table, where someone begins to read, and it sounded like this.

(At this point, Slide 10 is shown on the screen and behind the screen Dr. Ralph Fuhrman reads the citation for the Emerson Medal.)
(Slide 10—The Emerson Medal)

Ralph Fuhrman: To Howard Eugene Moses, upon the recommendation of the Awards Committee and approval by the Board of Control, The Charles Alvin Emerson Award for Meritorious service to the Federation of Sewage and Industrial Wastes Associations. I feel that there could not be a more deserving recipient than you, Ted. After you labored as one of the Committee of One Hundred, you took the assignment, so important in this formative period, of that of Secretary of the Federation. Through your 13 years in this office, you nurtured the infant organization from 600 affiliates to more than 3,000 individuals. Your guidance brought the Federation to the point of growth when it both needed and was able to support full time employees.

Those of us who realize that the token compensation you received for this work could not begin to pay you for your efforts, genuinely appreciate your most devoted service to this now international organization. At the same time, you served the Federation as secretary, you were my fellow director of the Pennsylvania Association and gave both that Association and the Federation Board of Control the benefit of your wise counsel. The Federation has honored you previously with Honorary Membership in 1943, and with the Kenneth Allen Award in 1945. It fills me with pride and satisfaction to confer on you, Ted, the Federation Award which bears my name.

GES: Who is it speaking and whose words was he reading?

Ted: That's Ralph Fuhrman—and he was quoting Charles Emerson's words.

(The curtain parts, and Fuhrman steps forward. Ted and Ralph exchange greetings)

GES: Yes, that's correct—Dr. Ralph E. Fuhrman, Exec. Sec. of the Federation.

Ralph was President of the Federation in 1951 and introduced Charlie Emerson, who actually read the citation you have just heard. As you know, Charlie is in the hospital and cannot be here to honor you tonight,* but he dictated a long letter in which he told about your work together in the Health Dept. Charlie asked me to ask you if you remembered when he sent you up to Erie County where there were about 250 cases of typhoid.

---

* Mr. Emerson died the following day.

And do you remember the time Fortenbaugh set fire to his pajamas in Kane, when he didn't know how to light the natural gas? And do you remember the mail box inspections and how Emerson checked on a crew of inspectors at a livery stable and found eight inspectors in eight single rigs lined up on the barn floor, waiting for the foreman to say go when his watch reached 8 o'clock? Do you remember that?

Ted: Yes! I do.

GES: I also bring you greetings from another Federation friend of yours, W.H. (Pete) Wisely, former Secretary of the Federation, who is now in Norway. Pete asked me specifically the night before he left to give you his best wishes, and remind you of the time when you told one of your stories, up at the Nittany Lion Inn, on Pete. The story went like this— that Pete had been out with his daughter to get her an evening gown; it was a strapless gown, and when she came home with it, Mrs. Wisely said; she's too young for that and Pete said, if she is big enough to hold it up, she is big enough to wear it. I might tell you Ted, that that daughter is now grown and married and Pete is a grandfather.

Ted: I guess she is big enough now.

GES: As the years roll on you are honored by many organizations: Fuller Awardee and Life Member of AWWA, Life Member of ASCE, Member and Chairman of the Ohio River Valley Sanitation Commission; many others.

It is now 1953, you are standing on the platform at the graduation exercises at Dickinson College and you are about to receive another honor. . . .

(Slide 11—Honorary D.Sc. Degree)

With that Honorary Doctor's Degree went this citation:

"Howard Eugene Moses; son of a methodist minister, with a lifetime in the field of engineering, you have guarded the health of the people of Pennsylvania for nearly 50 years and have contributed immeasurably to the common good in your services to state, interstate and national agencies devoted to safeguarding public water supplies and to fight stream pollution. You have been often honored for your distinguished services, and your College has noted with approval your recognition by outstanding professional engineering societies. Nor is your Alma Mater unmindful of your years of devotion as a loyal alumnus, and on this 55th anniversary of your graduation is happy to honor you.

"Therefore, upon recommendation of the Faculty to the Board of Trustees and by its mandamus, I admit you to the degree of Doctor of Science, honoris causa, with all the rights, privileges and distinction thereunto appertaining, in token of which I present you with this diploma and cause you to be invested with the golden hood of Dickinson College, appropriate to your degree.

Signed William W. Edel, Pres."

(Slide 12—Ted Moses and daughter Gene at Dickinson College)

GES: Here's a picture taken at the exercises that day.

(Behin curtain is Mrs. Wilkinson.)

Mrs. F.M. Wilkinson: And believe me, I certainly was proud to be there with Doctor Moses.

Ges: And who is that Ted?

Ted: That's Gene!

GES: Yes, Ted, this is your daughter, Mrs. Francis Wilkinson—not all of your life has been spent in your office.

(Slide 13—cottage at Ocean City)

GES: Do you recognize this picture, Ted?

Ted: Yes!

Mr. Wilkinson: Here is where we spent many pleasant weekends and summer holidays. Do you remember your first airplane ride to view the results of an ice storm? You had flown with me many times but doggone if I could take you deep sea fishing.

GES: Who is it, Ted?

Ted: That's Wilkie!

(Curtain parts and Mr. Wilkinson steps forward.)

Wilkie: Glad to be with you Dad and believe me, I'm finding out lots of things about you I didn't know.

Ted: I haven't any secrets any more.

GES: Yes, Ted, your son-in-law Francis M. Wilkinson, of Wyomissing, Pa.

You have spent 55 years in Harrisburg, many of them in this house.

(Slide 14—Picture of present home.)

(Behind curtains stands Virgie Thompson)

Virgie Acy Thompson: In 1906, I came as a young girl and entered the Moses' home primarily to do odd jobs and to look after little daughter Gene. Gradually, Mr. Moses gave me more and more responsibility and after I graduated from high school, I took on the entire family as a full time responsibility.

Ges: Who is it Ted?

Ted: Oh! That's Virgie, the most wonderful, wonderful girl in the world.

Virgie: Congratulations Mr. Moses! I'm mighty glad to be here and take a part in this honor that is bestowed upon you.

Ted: I'm glad to see you Virgie. You want to come home soon; your pie baking clients are asking about you. They want some more of your pies.

GES: Yes, Ted, this is Virgie Acy Thompson, still an important part of your household who must be given credit for a job well done. And now Ted—you've had a long and useful career, and there's much left yet ahead of you. And for that future, I cannot give you a Hazel Bishop lipstick,* or a movie and sound recording of these proceedings.

The committee does have for you, however, a small viewer and the slides which have been shown here tonight. And you will receive some of the pictures taken here tonight. Also you will receive this book containing the script of the show. And one more thing . . .

(Spotlight on curtains)

GES: Here are two more people you know . . .

(Out step presidents of two Pennsylvania associations—Vance Rigling and Gordie Wiest.)

GES: They have a gift for you—a plaque containing the two keys of the two organizations which honor you tonight—and the inscription reads:

"Howard Eugene (Ted) Moses—In your honor and in order to express our appreciation, we the Penna. Water Works Operators Assn. and the Penna. Sewage and Industrial Wastes Assn. present this plaque to you symbolizing your efforts in founding our associations and in furthering the interest in their growth and expansion."

(Presidents hand over plaque and shake hands)

Ted: Thanks so much! If you keep on, I'll have to cry for you or something like that.

GES: Sit down a minute Ted and then we're through. I want to take this opportunity in conclusion to thank all of those who participated in this affair, in helping get it ready; in particular I want to thank Al Estrada, Deac Matter, Bernie Bush, who could not be here, and Roy Kountz of the special committee, who conceived this tribute and who permitted me to participate in it and who furnished the information from which I wrote the script.

I also want to thank, for myself and I'm sure for you, the others to whom some credit is due; the boys who worked behind the scenes, Russ Kluck, Johnnie Yenchko, Jack Nesbitt, and Bob Thomas, all of whom helped.

This is your life, Ted Moses, and this has been Ted Moses Night. I know that there are many of your friends out there waiting to shake your hand, to thank you for a job well done and to wish you continued health and activity, just as I do; and so I give you over to this audience of well wishers. My thanks to you TED for being such a grand guy!

Ladies and Gentlemen—I give you Ted Moses!

(Applause)

(At this point, after a tremendous standing ovation, Mr. Moses took the microphone and expressed simply and movingly his sincere thanks, not only for the demonstration of affection in which he is held, but also for the many pleasant associations he had had over a long period of service to his state, his profession and his friends. Unfortunately those remarks were not recorded.)

Aug. 23, 1955

---

* Sponsor of the "This is Your Life" TV Show. Permission to do the "Ted Moses Show" was obtained from Ralph Edwards, who asked that his sponsor be mentioned.

H.E. Moses Memorandum Report
To: Dr. Mattison
Re: FSWA Mtg. Oct. 1955

Memorandum report October 17, 1955

TO: DR. MATTISON

FROM: H.E. MOSES, DIRECTOR
BUREAU OF SANITARY ENGINEERING

Re: Federation of Sewage and Industrial Wastes Associations Meeting, Atlantic City, October 10-13, 1955

It was my good fortune to attend the above mentioned meeting, which turned out to be one of the most successful held by this organization, which is international in its nature, and also the largest ever had with a total attendance of over 1300 persons. Ambassador Hotel was the headquarters with the overflow housed in the Ritz-Carlton Hotel nearby. Currently, with the meeting there was an extremely large exhibit of machinery and equipment used in treatment works for sewage and industrial wastes. This added greatly to the benefit of those attending the meeting, particularly since the group was mainly composed of engineers, chemists, scientists, and operators practicing chiefly within the field of sanitation.

As is usual in the Federation meetings, the subjects scheduled related specifically to matters of stream pollution control, and presenting them and participating in the discussions were men expert in the field and leaders of their profession. Along with these discussions were descriptions of installations made in various parts of the country, which should have a marked effect in extending the area of stream pollution control.

One topic of much interest to Pennsylvania was a description of the Wilmington, Delaware sewerage system, which indicated that Pennsylvania's neighboring state in the Delaware Basin is keeping pace with the stream control measures put into execution by Pennsylvania. Consideration was given to the acid mine drainage problem; also to modern methods of sludge disposal, which is always an important problem in treatment plants. Another paper described High-Rate trickling filters in Germany. This was presented by Adolf Rumpf, one of five German scientists who are visiting in this country to observe American methods of handling harmful wastes. Another up-to-date topic was the use of radioactive isotopes in determining flow patterns in waste treatment lagoons. There was a symposium on fringe-area sanitation, a problem of much concern to this Department at the present time. Still another was the design and operation of sewage treatment plants along the Turnpikes, the first of which were installed in the western section of the Pennsylvania Turnpike.

On Thursday, October 13, there were two Forums, one pertaining to Sewerage and the other to Industrial Wastes. Mr. Ralph M. Heister, Principal Sanitary Engineer of this Department, presented a paper in the Industrial Wastes session on a topic relevant to that particular field of engineering.

Because of the country-wide interest in federal stream pollution control legislation, a meeting was held at Atlantic City on Tuesday, October 11. This was called by Milton P. Adams, Executive Secretary of the Michigan Water Resources Committee. The other four members of the Committee comprised David B. Lee, Florida; Vinton W. Bacon, California; Edward J. Cleary, ORSANCO; and James H. Allen, INCODEL. It was anticipated that this would be a small group but it turned out to be comprised of probably 40 men from all over the United States. It was reported that some 32 states had representatives present, either members of the State Health Department or of interstate agencies.

Mr. Adams was appointed Chairman, and Mr. Morris M. Cohn of New York State acted as Secretary. The meeting lasted for two hours, and the discussion was quite spirited but was held by Mr. Adams to the objective of the meeting, namely, to discuss the various aspects of the Federal legislation, particularly Senate Bill 890, which, during the last Session of Congress, passed the Senate and was sent to the Committee on Public Works in the House of Representatives, which committee reported it out, accompanied by a report calling attention to some of the salient and much discussed features of the bill. It did not come to a vote, and it is anticipated that the matter will again be presented at the next Session of Congress for perhaps further hearing and final action.

It appeared that the bill as it now stands after various amendments is not too far from being acceptable generally throughout the country. However, there are still some points in question, which were the subject of discussion at this meeting, and this finally resolved itself into detailed consideration of certain proposed further amendments to two principal sections of the bill. One of these relates to the procedure by which the Surgeon General of USPHS will follow in the event of a case of stream pollution. There was presented for consideration of the group certain amendments offered by Mr. Fred Zimmerman, representing the New York Joint Legislative Committee on Interstate Cooperation. These involved a discussion of various features of Sections 9(b) and 6(a), both of which in the original bill seemed to be generally unsatisfactory to the country at large, the first because of the lack of proper notification by the Surgeon General to regulatory agencies of his intention to place an order upon offenders, and the second as to the composition of the Water Pollution Control Advisory Board. In the first instance, the purport of Mr. Zimmerman's suggested amendment was to extend the time relating to the giving of notices, and the second was to more nearly balance representation of Federal and State agencies in the Advisory Board.

Finally the group expressed it as the consensus of opinion that the proposed amendments offered by Mr. Zimmerman should be included in the amended bill (S.890), and that appropriate steps should be taken when Congress reconvenes to procure such inclusion. Another point discussed and rather generally agreed to was the desirability of having the bill sent back to the House Committee on Public Works for further hearing.

Federation Affairs*

At its 33rd Annual Meeting in Philadelphia, October 2-6, the Water Pollution Control Federation adopted the following statement of policy regarding national water pollution problems. The Federation has a real responsibility to the public as well as to its membership and it believes that by aggressively pursuing the objectives of this statement the nation's health and welfare can be benefited.

We urge you to read, weigh, and use this statement carefully to see where each of us can best fit our talents toward the fulfillment of these objectives.

STATEMENT ON WATER POLLUTION CONTROL
Adopted by the Board of Control of the
Water Pollution Control Federation, October 6, 1960

Pollution of the nation's watercourses, ground waters, and coastal waters is a continuing hazard to health, comfort, safety, and economic welfare. While considerable progress has been made in pollution control by many municipalities and industries, many water resources areas are being degraded, impaired, and damaged by such discharges, and they will be further adversely affected by the amount and pattern of population growth, industrial expansion, and technological advancements.

To Assure the Conservation and Protection of the Nation's Water Resources, the Water Pollution Control Federation Believes:

1. That the discharge of pollutional wastes into the waterways of the nation should be controlled.
2. That the type and extent of treatment and control for any specific situation must be determined after consideration of the technical factors involved.
3. That the responsibilities for the adequate treatment and control of wastes to overcome pollution must be shared individually and jointly by industry and local, state, and federal governments.
4. That basic and applied research by competent personnel must be encouraged by broad mutual effort to develop new knowledge that will solve water pollution problems.
5. That the administration of pollution control must be firm, effective, and equitable.
6. That the administration of state and interstate pollution control programs should remain in the hands of state and interstate water pollution control agencies which must be supported by increased budgets and adequately staffed by well-trained and compensated engineers, scientists, and other personnel.
7. That the primary objective of pollution control is the protection of the public health, with other objectives adding impelling reasons for water resources protection.

---
* Reprinted from *Jour. Water Poll. Control Fed.,* **32,** 1257 (1960).

8. That federal activity in water pollution control should be administered by the Public Health Service, the organization best fitted to perform these functions by virtue of its long experience and close cooperation with state health departments and state and interstate water pollution control agencies.
9. That the public must be made fully aware of the hazards of pollution and of the workable means for control, so that it will sponsor and support construction and proper operation of all necessary facilities.
10. That mandatory certification or licensing of better-trained and compensated operating personnel is the best ultimate means for assuring the most effective operation and maintenance of pollution control facilities.
11. That standards for radiation hazards in water pollution control should be primarily in the interest of the protection of the public health.
12. That the control of toxic and exotic chemicals should be exercised, to the maximum extent practicable, at the source in order to prevent problems in water pollution control.
13. That federal, state, and local fiscal laws and practices should be devised and modified to assure the most economical and effective means for financing the construction, operation, and upgrading of wastewater treatment works.

Note: Morris Cohn, editor of Wastes Engineering, *who served as a Federation Constitutional Committee chairman, was one of the first to publicize the Federation's new policy on water pollution control. The full text of the statement was published as part of an article in the November, 1960 issue of* Wastes Engineering *under the title, "This We Believe."*

The WPCF policy statement was revised as needed to keep abreast of expansion and changes in the water pollution control field. The latest version was adopted by the Board of Control on October 10, 1974, and is presented here:

STATEMENT OF POLICY ON WATER POLLUTION CONTROL IN THE UNITED STATES

Water pollution means water quality damage and consequent interference with beneficial use of a vital resource—*clean water.*

Pollution of the Nation's inland surface waters, coastal waters, and groundwaters is a continuing threat to the national health, aesthetic enjoyment, safety, and economic welfare. National survival, in terms of future urban, industrial, and commercial growth and prosperity, dictates the protection of all water resources from any acts, such as the discharging of harmful substances which cause unreasonable impairment of water quality and adversely affect their highest level of usefulness. While considerable progress has been made in pollution control by municipalities and industries, many water resources are being degraded, impaired, and damaged by such discharges and acts, and they will be further adversely affected by the degree and pattern of population growth, industrial processing, commercial expansion, chemical usages, agricultural developments, and other technological advancements.

The Water Pollution Control Federation is pledged to provide leadership and guidance to all constructive efforts that contribute to the control of water pollution. Its pledge is summarized by the following points.

1. The objectives of water pollution control must include preservation of high quality waters for protection of public health; for industrial, agricultural, and recreational uses; for fish and wildlife propagation; and for the maintenance of an aesthetically desirable environment.
2. The discharge of all wastewater into the waters of the nation must be controlled in a rational manner. Such regulatory control must be based not only on considerations of specific wastewater discharge characteristics but also on additional factors including discharge location, physical, chemical, and biological characteristics of the receiving waters, defined beneficial uses, and appropriate water quality criteria in order to provide adequate protection of the beneficial uses of the environment. After a facility is in operation additional ecological and environmental studies should be carried out to determine the effectiveness of the facility and the need for future modification of the facility.
3. The responsibilities for the adequate treatment and control of wastewater must be assumed individually and jointly by industry and local, state, interstate, and federal governments.
4. The administration of water pollution control must be firm and effective and should remain in the hands of state and interstate water pollution control agencies. Regulatory agencies must be supported by adequate budgets and fully staffed by competent engineers, scientists, and supporting personnel.
5. Federal, state, and local laws and practices must reflect the changing needs in order to obtain and maintain the most economical and effective means for financing the construction, management, operation, and maintenance of wastewater collection systems and treatment works.
6. The public must be made fully aware of the consequences of water pollution and the costs of its control. Only in this way can the public be prepared to sponsor and support sound water pollution control measures.
7. Basic and applied research by competent personnel must be encouraged by broad efforts to develop new knowledge that will solve water pollution problems.
8. Wastewater represents an increasing fraction of the Nation's total water resources and should be reclaimed for beneficial reuse. To this end the development and application of methods for wastewater reclamation must be accelerated.
9. Mandatory certification or licensing of adequately trained and properly compensated personnel must be encouraged as a requirement for maximum effectiveness of treatment facilities.

Letter from Walter A. Sperry to G.E. Symons

Thursday, Aug. 14, 1969

Dear Friend George—

"Your interesting letter of August 8th stirred a lot of memories. With this I am enclosing the 1927-42 Silver Anniversary Report of Progress of the then Central States Association. Thinking that it ought not to be buried in my library but filed where it might be more available, you may keep it.

At age 76 I retired Jan. 1, 1958, after 51 years in the water and sewage field; I had helped start and operate the famous Columbus, Ohio filter plant. Was at Grand Rapids 17 years, having started and operated that plant and then started Prof. Hoad's Saginaw Plant long enough to teach every one their jobs. In all some 20 odd years. After a year with the State of Michigan as the first Secretary of their Stream Control Comm., which Milt Adams so efficiently ran, I came to Aurora for the next 28 years.

During that time and for how many years I forget I was editor of the *Windmill* and also for ten years I had a 4-page column of paragraphs in the *Journal*. It was during that time that I planned the enclosed booklet

One special privilege was the training of Lloyd Billings for 17 years who made his mark and here at Aurora I trained Willard Pfeifer for 28 years, who took my place and is giving a brilliant account of himself.

Pardon this bit of remembering but I loved every minute of the 51 years. I will be 87 this September and my eyes are such that I can not read most technical journals due to gloss paper and light inking and must use a strong lens.

Hope the enclosed proves useful and so with kind memories and well wishing I am,

Cordially,
Walter A. Sperry

History of Selected Constitutional Committees

(Committees whose chairmen are *ex officio*, such as Executive, Policy Advisory, Nomination, and Honorary Membership, are not included.)

Finance Committee—Established as a special committee (Finance Advisory) before it became a constitutional committee in 1949. Chairmen: W.J. Orchard, 1941-1961; R.F. Orth, 1961-1966; F.H. Miller, 1966-1967.

Organization Committee—Included as constitutional in the Constitution and Bylaws of 1941; discontinued by amendment to the Constitution and Bylaws September 27, 1964. Chairmen: E. Boyce, 1941-1952; H.W. Streeter, 1952-1954; S.G. Hess, Jr., 1954-1956; J.W. Wakefield, 1956-1961; B.A. Poole, 1961-1963; W.Q. Kehr, 1963-1964.

Constitution and Bylaws Committee—Became a constitutional committee by amendment to the Constitution and Bylaws September 27, 1964; before that time it was a special committee. Chairmen: A.J. Fischer, 1964-1967; W.Q. Kehr, 1967-1969; E.R. Howard, 1969-1970; J.L. Robinson, 1971-1972; C.A. Parthum, 1972-1977.

Publications Committee—Shown as a constitutional committee in the Constitution and Bylaws of 1941. It had responsibility for the technical programs of the conferences until 1963-'64. Its name was changed to Publications and Program by amendment of the Constitution and Bylaws October 7, 1956, and back to Publications October 6, 1963. Chairmen: F.W. Gilcreas, 1941-1954; R. Eliassen, 1954-1956; G.E. Symons, 1956-1961; P.D. Haney, 1961-1963; C.C. Larson, 1963-1968; K.S. Watson, 1968-1973; P.D. Haney, 1973-1977.

Program Committee—Formed as a constitutional committee by amendment to the Constitution and Bylaws October 6, 1963. Chairmen: P.D. Haney, 1963-1966; L.F. Oeming, 1966-1969; J.H. Robertson, 1969-1974; P.A. Krenkel, 1974-1977.

Research Committee—Shown as constitutional in the Constitution and Bylaws of 1941. Chairmen: W. Rudolfs, 1941-1953; H. Heukelekian, 1953-1961; D.A. Okum, 1961-1966; C.M. Weiss, 1966-1971; R.S. Engelbrecht, 1971-1975; R.S. Englebrecht and F.G. Pohland (co-chairmen), 1975-1976; F.G. Pohland, 1976-1977.

Standard Methods Committee—Became a constitutional committee by amendment to the Constitution and Bylaws December 15, 1953. It had existed as a special committee before that time. Chairmen: G.P. Edwards, 1953-1960; R.S. Ingols, 1960-1965; R.D. Hoak, 1965-1970; M.C. Rand, 1970-1975; D. Jenkins, 1975-1977.

Technical Practice Committee—Shown as a constitutional committee in the Constitution and Bylaws of 1941 with the name Sewage Works Practice. This name was changed to Sewage and Industrial Wastes Practice by amendment to the Constitution and Bylaws October 8, 1950, and to Technical Practice by amendment to the Constitution and Bylaws October 6, 1963. Chairmen: M.M. Cohn, 1941-1961; D. E. Bloodgood, 1961-1966; V.W. Langworthy, 1966-1971; W.F. Garrison, 1971-1976; H.G. Schwartz, Jr., 1976-1977.

Industrial Wastes Committee—Made constitutional by amendment to the Constitution and Bylaws October 7, 1956. Chairmen: R.W. Hess, 1956-1958; K.S. Watson, 1958-1963; J.F. Byrd, 1963-1968; R. Rocheleau, 1968-1971; R.D. Sadow, 1971-1975; G.N. McDermott and R.D. Eadow (co-chairmen), 1975-1976; G.N. McDermott, 1976-1977.

Government Affairs Committee—Made a constitutional committee by amendment to the Constitution and Bylaws October 9, 1969. Prior to that time it existed as a special committee with names Legislative Analysis and later as Legislative. Chairmen: V.G. Wagner, 1969-1973; E.J. Newbould, 1973-1977.

WPCF-AWWA Joint Resolution on Water Reuse, 1973

Promotion of water reuse in the 1960's led to serious questions concerning adequacy of technology to evaluate the safety of using reclaimed wastewater in drinking water systems. As a result of discussions in water reuse committees of WPCF and AWWA, both organizations agreed to the formation of a joint committee to develop a position on this subject. The joint committee recommended and both organizations approved in 1973 the following resolution:

Joint Resolution of the American Water Works Association and the Water Pollution Control Federation on Potable Reuse of Water

WHEREAS: Ever-greater amounts of treated wastewaters are being discharged to the waters of the nation and constitute an increasing proportion of many existing water supplies, and

WHEREAS: More and more proposals are being made to introduce reclaimed wastewaters directly into various elements of domestic water supply systems, and

WHEREAS: The sound management of our total available water resources must include consideration of the potential use of properly treated wastewaters as part of drinking water supplies, and

WHEREAS: There is insufficient scientific information about acute and long-term effects on man's health resulting from such use of wastewaters, and

WHEREAS: Fail-safe technology to assure the removal of all potentially harmful substances from wastewaters is not available,

NOW THEREFORE BE IT RESOLVED: That the American Water Works Association and the Water Pollution Control Federation do hereby urge the Federal Government to support an immediate and sustained multidisciplinary national effort to provide the scientific knowledge and technology relative to the reuse of water for drinking purposes in order to assure the full protection of the public health.

WPCF GROWTH TRENDS
Net Worth

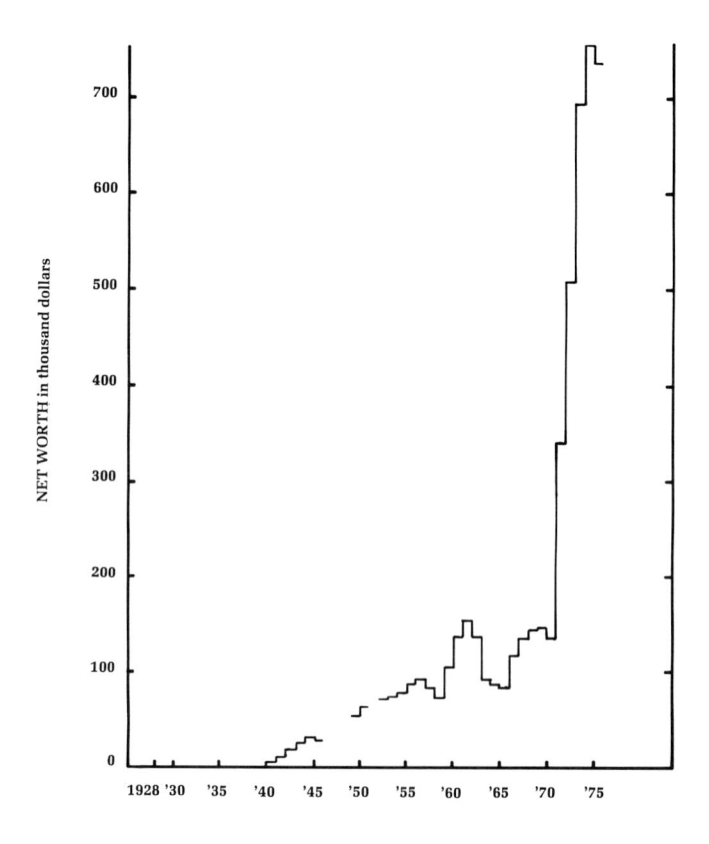

WPCF GROWTH TRENDS
Annual Budget

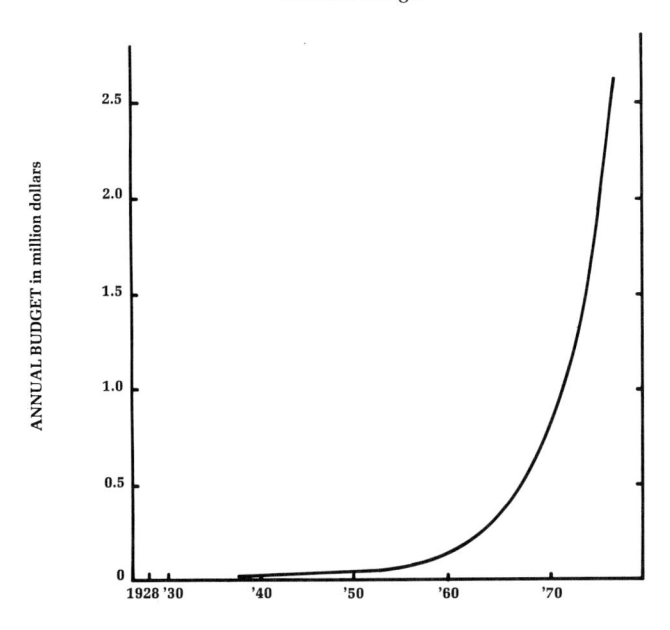

WPCF GROWTH TRENDS
Membership and Staff

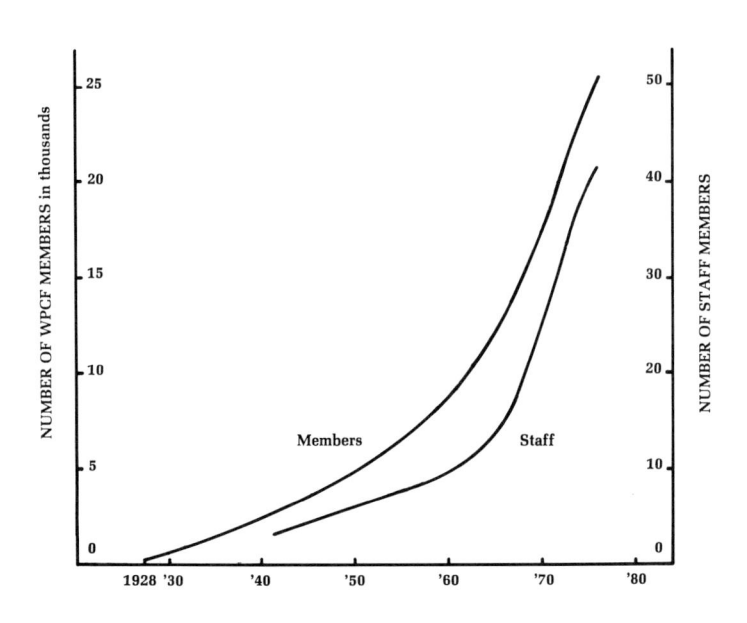

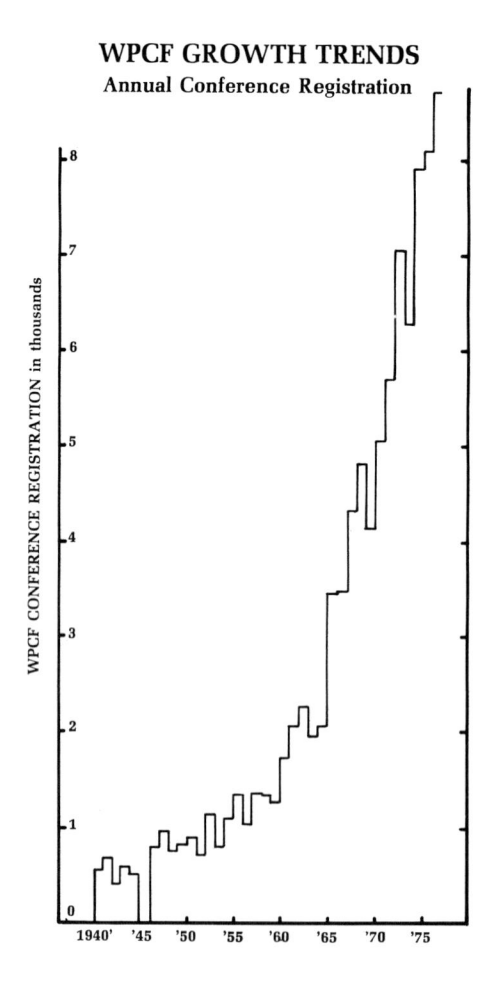

WPCF GROWTH TRENDS
Annual Conference Registration

WPCF GROWTH TRENDS
Exhibitors and Space Sold at Annual Conferences

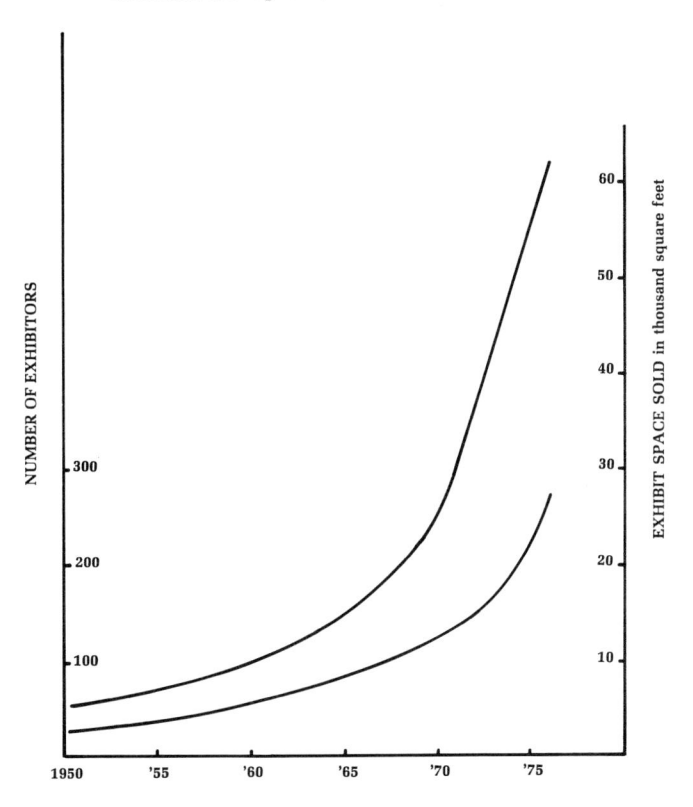